MAGNETIC HYSTERESIS

IEEE Press
445 Hoes Lane, P.O. Box 1331
Piscataway, NJ 08855-1331

IEEE Magnetics Society, *Sponsor*

Cover design: William T. Donnelly, *WT Design*

Technical Reviewers

Stanley H. Charap, *Carnegie Mellon University*
John Oti, *Western Digital*

MAGNETIC HYSTERESIS

Edward Della Torre
The George Washington University

IEEE PRESS

IEEE MAGNETICS SOCIETY, *SPONSOR*

The Institute of Electrical and Electronics Engineers, Inc., New York

This book and other books may be purchased at a discount from the publisher when ordered in bulk quantities. Contact:

IEEE Press Marketing
Attn: Special Sales
Piscataway, NJ 08855-1331
Fax: (732) 981-9334

For more information about IEEE PRESS products, visit the IEEE Home Page: http://www.ieee.org/

Printed in the United States of America

10 9 8 7 6 5 4 3 2 1

ISBN 0-7803-4719-6
IEEE Order Number: PC5766

Library of Congress Cataloging-in-Publication Data

Della Torre, Edward 1934-
Magnetic Hysteresis / Edward Della Torre.
p. cm.
Includes bibliographical references and index.
ISBN 0-7803-4719-6
1. Hysteresis. I. Title.
QC754.2.H9T67 1999
538'.3- -dc21 98-46940
CIP

To the memory of
Charles V. Longo

CONTENTS

PREFACE

The modeling of magnetic materials can be performed at various levels of resolution. The highest level of resolution is the atomic level. At this level, one can use quantum mechanics to understand the basic processes involved. The next step down in resolution is the micromagnetic level, where the magnetization is a continuous function of position. At a still lower level of resolution, one uses the domain level of modeling, where the material is divided into uniformly magnetized domains separated by domain walls of zero thickness. Finally at the lowest resolution, the nonlinear medium level, the magnetization is the average of many domains, and the physical nature of their formation is ignored. In this last level, the medium is characterized by an input/output relationship.

Preisach modeling is a mathematical tool that has been used principally at the nonlinear medium level, but it can also give some insight at all the levels. Its effectiveness in describing magnetic materials is due to its ability to have a behavior when the applied field is increasing which is different from its behavior when the applied field is decreasing. It is thus able to describe minor loops and other complex magnetizing processes. The classical Preisach model is limited by the congruency property and the deletion property, neither of which is possessed by magnetic materials. Although these limitations could be removed using a phenomenological approach, this book relies on physical reasoning as much as possible to make necessary modifications. This practice usually results in simpler models that give physical insight into the processes of interest. Although these modifications have been shown to be robust, the book uses physical reasoning rather than mathematical rigor to justify its derivations.

In Chapter 1, the physics of magnetization processes is briefly summarized. Chapter 2 summarizes the classical Preisach model, which is the basis for the statistical analysis used in modeling hysteresis. However, since it cannot describe many of the subtleties in the behavior of magnetic materials, modifications based upon physical reasoning are presented in the subsequent chapters. In particular, the concept of reversible magnetization is discussed in Chapter 3. Accurate behavior of the susceptibility is obtained by a magnetization-dependent reversible component, called the DOK model. This is further improved by adding a more complex state-dependent reversible component, called the CMH model. As shown in Chapter 4, the congruency limitation can be removed by means of an output-dependent model, such as the moving model or the product model. Including either accommodation, aftereffect, or both in the model, as shown in Chapter 5, removes the deletion property. Even with all these modifications, the resulting model is still a scalar model, so in Chapter 6, we discuss methods of generalizing it to a vector model.

Some applications are discussed in Chapter 7. First, since the model is essentially a magnetostatic model, this chapter presents two brief extensions to dynamics. These extensions include the effect of eddy currents on the magnetization, the effect of the accommodation model on the pulsed behavior, and the effect of the moving model and the accommodation model on noise. Another extension is the development of a magnetostriction model. Finally, the development of an inverse model, which would be useful in control applications, is discussed.

I hope that this book is useful in showing how the Preisach model can be extended to describe accurately a wide range of magnetic phenomena. Although the discussion is limited to magnetic phenomena, it can give deep insight into the analysis of hysteretic many-body problems. The techniques presented here are general and can be applied to hysteresis problems in disciplines other than magnetism.

Edward Della Torre

ACKNOWLEDGMENTS

Ferenc Vajda deserves my special thanks. This book is the result of the many fruitful and stimulating discussions that we have had. Many of the numerous papers on which we had collaborated form an important part of this book. I would like to single him out for his earlier help, insight and encouragement.

I also thank the following students, who attended a course in which I used the manuscript of this book as a text: Jason Eicke, Luis Lopez-Diaz, Jie Lou, Ann Reimers, and Pattana Rugkwamsook. I am grateful to Lawrence H. Bennett, who has been a constant source of advice and encouragement. Also Michael Donahue, Robert McMichael, and Lydon Swartzendruber deserve my thanks. My many colleagues, too numerous to mention, with whom discussions resulted in a rich exchange of ideas, also are acknowledged here with thanks.

I also thank my wife, Sonia, who read this manuscript and made many helpful suggestions as it progressed.

Edward Della Torre

CHAPTER 1

PHYSICS OF MAGNETISM

1.1 INTRODUCTION

The aim of this book is to characterize the magnetization that results in a material when a magnetic field is applied. This magnetization can vary spatially because of the geometry of the applied field. The models presented in this book will compute this variation accurately, provided the scale is not too small. In the case of particulate media, the computation cells must be large enough to encompass a sufficient number of basic magnetic entities to ensure that the deviation from the mean number of particles is a small fraction of the number of particles in that cell. In the case of continuous media, the computation cells must be large enough to encompass many inclusions. The study of magnetization on a smaller scale, known as *micromagnetics*, is beyond the scope of this book. Nevertheless, we will see that it is possible to have computation cells as small as the order of micrometers.

This book presents a study of magnetic hysteresis based on physical principles, rather than simply on the mathematical curve-fitting of observed data. It is hoped that the use of this method will permit the description of the observed data with fewer parameters for the same accuracy, and also perhaps that some physical insight into the processes involved will be obtained. This chapter reviews the physics underlying the magnetic processes that exhibit hysteresis only in sufficient detail to summarize the theory behind hysteresis modeling; it is not intended as an introduction to magnetic phenomena.

This chapter's discussion begins at the atomic level, where the behavior of the magnetization is governed by quantum mechanics. This analysis will result in a methodology for computing magnetization patterns called micromagnetism. For a more detailed discussion of the physics involved, the reader is referred to the excellent books by Morrish [1] and Chikazumi [2].

Since micromagnetic problems involve hysteresis, there are many possible solutions for a given applied field. The particular solution that is appropriate depends on the history of the magnetizing process. We view the magnetizing process of hysteretic media as a many-body problem with hysteresis. In this chapter, we start by reviewing some physical principles of magnetic material behavior as a basis for developing models for behavior. Special techniques are devised in future chapters to handle this problem mathematically. The Preisach and Preisach-type models, introduced in the next chapter, form the basic framework for this mathematics. The discussion presented relies on physical principles, and we will not discuss the derived equations with mathematical rigor. There are excellent mathematical books addressing this subject, including those by Visintin [3] and by Brokate and Sprekels [4]. In subsequent chapters, when we modify the Preisach model so that it can describe accurately phenomena observed in magnetic materials, we will see all these physical insights and techniques.

1.2 DIAMAGNETISM AND PARAMAGNETISM

Both diamagnetic and paramagnetic materials have very weak magnetic properties at room temperature; neither kind displays hysteresis. Diamagnetism occurs in materials consisting of atoms with no net magnetic moment. The application of a magnetic field induces a moment in the atom that, by Lenz's law, opposes the applied field. This leads to a relative permeability for the medium that is slightly less than unity.

Paramagnetic materials, on the other hand, have a relative permeability that is slightly greater than 1. They may be in any material phase, and they consist of molecules that have a magnetic moment whose magnitude is constant. In the presence of an applied field, such a moment will experience a torque tending to align it with the field. At a temperature of absolute zero, the electrons or atoms with a magnetic moment in assembly would align themselves with the magnetic field. This would produce a net magnetization, or magnetic moment per unit volume, equal to the product of their moment and their density. This is the maximum magnetization that can be achieved with this electron concentration, and thus it will be called the *saturation magnetization* M_S. Atoms possess a magnetic moment that is an integer number of Bohr magnetons. The magnetic moment of an electron, m_B, is one Bohr magneton, which in SI units is 0.9274×10^{-23} A-m². We note that the permeability of free space μ_0, and Boltzmann's constant, k, are in SI units $4\pi \times 10^{-7}$ and 1.3803×10^{-23} J/mole-deg, respectively.

Paramagnetic behavior occurs when these atoms form a reasonably dilute electron gas. At temperatures above absolute zero, for normal applied field strengths, thermal agitation will prevent them from completely aligning with that field. Let us define B as the applied magnetic flux density, and T as the absolute temperature. Then if we define the *Langevin function* by

$$L(\xi) = \coth \xi - \frac{1}{\xi}, \tag{1.1}$$

then the magnetization is proportional to the Langevin function, so that

$$M = M_S \, L(\xi), \tag{1.2}$$

where

$$\xi = \frac{\mu_0 g J m_B H}{kT} = \frac{\mu_0 m H}{kT}. \tag{1.3}$$

Here the moment of the atom, m, is the product of g, the gyromagnetic ratio, J the angular momentum quantum number, and m_B the Bohr magneton. It can be shown that the distribution of magnetic moments obeys Maxwell–Boltzmann statistics [5]. Figure 1.1 shows a plot of the Langevin function and its derivative. It is seen that for small ξ the function is linear with slope 1/3 and saturates at unity for large ξ.

The susceptibility of the gas, the derivative of the magnetization with respect to the applied field, is given by

$$\chi(H) = \frac{dM}{dH} = \frac{M_S \xi}{H}\left[\frac{1}{\xi^2} - \mathrm{csch}^2(\xi)\right]. \tag{1.4}$$

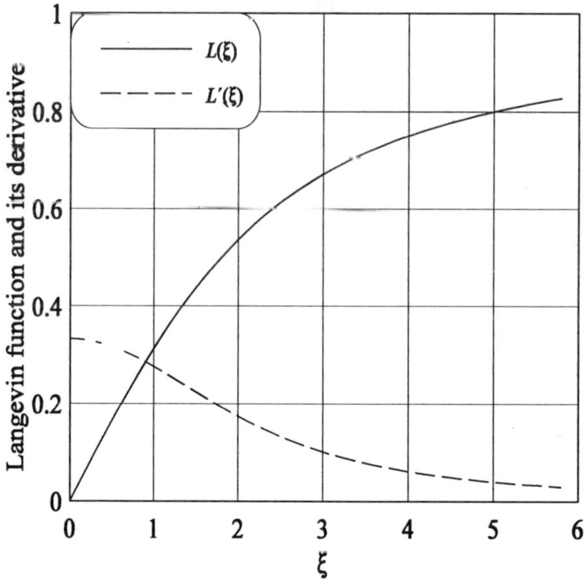

Figure 1.1 Langevin function (solid line) and its derivative (dashed line).

For small ξ, the quantity in brackets approaches 1/3. Thus, when the applied fields are small, the susceptibility, χ_0 is given by

$$\chi_0 = \frac{M_S \xi}{3H} = \frac{\mu_0 m M_S}{3kT}. \tag{1.5}$$

The small field susceptibility as a function of temperature is shown in Fig. 1.2.

At room temperature, the argument of the Langevin function is very small, and this effect is very weak; that is, the misalignment due to thermal motions is much greater than the effect of the applied field. Thus, this effect is not significant in the description of hysteresis; however, when discussing ferromagnetic materials, we will see that their behaviors above the Curie temperature are similar, except that the susceptibility diverges at the Curie temperature rather than at absolute zero.

The previous analysis did not include quantization effects. Since the magnetic moment can vary only in integer multiples, the Langevin function must be replaced by the Brillouin function, $B_J(\xi)$. The *Brillouin function* is defined by

$$B_J(\xi) = \frac{2J+1}{2J}\coth\frac{2J+1}{2J}\xi - \frac{1}{2J}\coth\frac{\xi}{2J}. \tag{1.6}$$

Thus, the magnetization $M(T)$ at temperature T, is given by

$$M(T) = N m_B g J B_J(\xi), \tag{1.7}$$

where, N is the number of atoms per unit volume, g is 0.5 for the electron, and J, an integer, is the angular momentum quantum number. The Brillouin function is zero if ξ is zero, and approaches one if ξ becomes large, as seen in Fig. 1.3. Therefore, from (1.7) we have

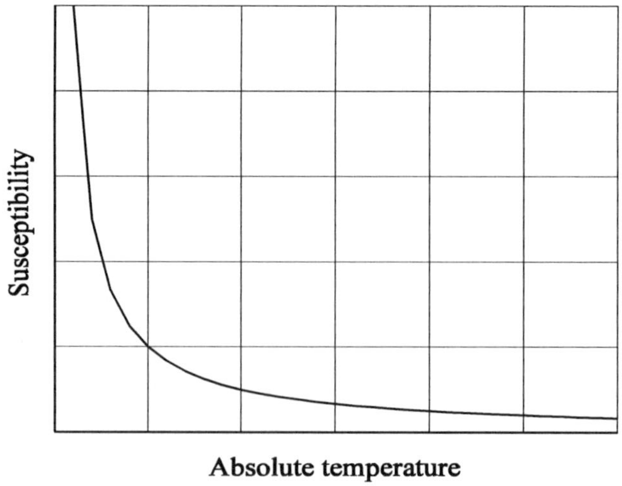

Figure 1.2 Paramagnetic susceptibility as a function of temperature.

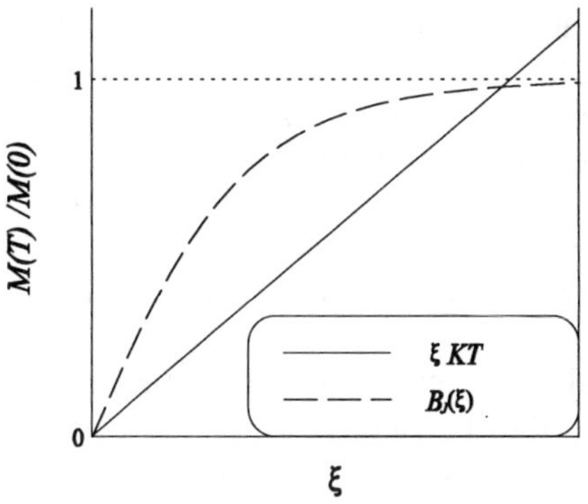

Figure 1.3 Plot of B_J and the linear function, ξKT as a function of ξ for $J = 1$.

$$M(0) = N m_B g J, \tag{1.8}$$

and so

$$\frac{M(T)}{M(0)} = B_J(\xi). \tag{1.9}$$

For small ξ, the Brillouin function is given approximately by

$$B_J(\xi) \approx \frac{J+1}{3J}\xi. \tag{1.10}$$

1.3 FERRO-, ANTIFERRO- AND FERRIMAGNETIC MATERIALS

In accordance with the Pauli exclusion principle, electrons obey Fermi–Dirac statistics; that is, only one electron can occupy a discrete quantum state at a time. When atoms are placed close together as they are in a crystal, the electron wave functions of adjacent atoms may overlap. Here, it is found that given a certain direction of magnetization for one atom, the energy of the second atom is higher for one direction of magnetization than the other. This difference in energy between the two states is called *exchange energy.* Furthermore, when the parallel magnetization is the lower energy state, the exchange is said to be *ferromagnetic*, but when the antiparallel magnetization is the lower energy state, the exchange is said to be *antiferromagnetic*. In ferromagnetic materials, this energy is very large and causes adjacent atoms to be magnetized in essentially the same direction at normal temperatures. Pure metal crystals of only three elements, iron, nickel, and cobalt, are ferromagnetic.

Since the electron wave functions are very localized, the overlap of wave functions between adjacent atoms decreases very quickly to zero as a function of the distance between them. Thus, exchange energy is usually limited to nearest neighbors. Sometimes the intervening atoms in a compound can act as a medium so that more distant atoms can be exchange coupled. Here, the resulting exchange is called *superexchange*. This, can also be either ferromagnetic or antiferromagnetic. Thus, compounds such as chromium dioxide can also be ferromagnetic.

The effect of exchange energy can be accounted for by an equivalent *exchange field*. Thus, the field, H, that an atomic moment experiences is given by

$$H = H_A + N_W M, \tag{1.11}$$

where H_A is the applied field, N_W is the molecular field constant, and $N_W M$ is the exchange field. Substituting this into (1.3), one sees that ξ is now given by

$$\xi = \frac{\mu_0 g J m_B [H + N_W M(T)]}{kT}. \tag{1.12}$$

The remanence is obtained by setting H equal to zero in this equation and solving for $M(T)$. Thus,

$$M(T) = \frac{\xi kT}{\mu_0 g J m_B N_W}, \tag{1.13}$$

and we can use (1.8) to write this as follows:

$$\frac{M(T)}{M(0)} = \frac{\xi kT}{\mu_0 g^2 m_B^2 J^2 N N_W} = \xi KT. \tag{1.14}$$

Since this must also be equal to the Brillouin function, we can obtain a graphical solution by plotting the two functions on the same graph, as illustrated in Fig. 1.3. For low temperatures, the slope of (1.14) is very small, so the intersection occurs at large values of ξ, and thus normalized magnetization approaches unity. As the temperature increases, the slope also increases, and thus, the magnetization decreases.

At the Curie temperature, Θ, the slopes of (1.14) and that of the Brillouin function are equal. This intersection occurs at a point where both ξ and the magnetization are zero. The Curie temperature can be computed, since from (1.10), the slope of the Brillouin function is given by

$$\left.\frac{dB_J(\xi)}{d\xi}\right|_{\xi=0} = \frac{J+1}{3J} = K\Theta. \tag{1.15}$$

Thus, the Curie temperature is

$$\Theta = \frac{\mu_0 g^2 m_B^2 J(J+1) N N_W}{3k}. \tag{1.16}$$

Becker and Döring [6] computed the saturation magnetization as a function of temperature and the total angular momentum. A comparison with measured values for iron and nickel, as shown in Fig. 1.4, appears to be a good fit with theory if J is taken to be either 0.5 or 1.

Above the Curie temperature, the material acts as a paramagnetic medium with the susceptibility diverging at a temperature called the *Curie–Weiss temperature* rather than at absolute zero degrees. The latter temperature is close to the Curie temperature for most materials. This type of behavior occurs regardless of whether the material is single crystal or consists of many particles or grains that are larger than a certain critical size. However, for small particles or grains another effect occurs. We will show in Section 1.6 that if these grains are sufficiently small, they may have only two stable states separated by an energy barrier. It is then possible that at a temperature smaller than the Curie temperature, called the *blocking temperature*, the thermodynamic energy kT will become comparable to the barrier energy. In that case, the particles or grains can spontaneously reverse and the material no longer will appear to be ferromagnetic. Above the blocking temperature, it behaves like a paramagnetic material with grains that have moments much larger than the spin of a single electron. This type of behavior is called

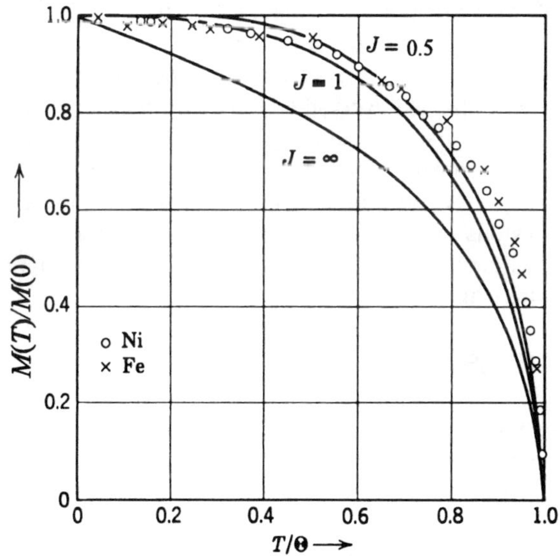

Figure 1.4 Temperature variation of saturation magnetization for atoms with different total angular momentum. [After Becker and Döring, 1939.]

superparamagnetism. As the temperature is raised from below, the material appears to lose its remanence and has a sudden large increase in its susceptibility. For a medium with a distribution of grain sizes, there is a distribution in energy barriers so that the blocking temperature is diffuse.

If the exchange energy is negative, it is convenient to think of the material as composed of two sublattices magnetized in the opposite directions. If the magnitude of the magnetization is the same for these two antiparallel sublattices, the net magnetization will be zero, and the material is said to be *antiferromagnetic* and appears to be nonmagnetic. On the other hand, if the magnitude differs, the material will have a net magnetization; such a material is said to be *ferrimagnetic.*

Ferrimagnetic materials usually have smaller saturation magnetization values than ferromagnetic materials, because the two sublattices have opposite magnetization. These materials are important, since they usually occur in ceramics that either are insulators or have very high resistivity. Such materials will support negligible eddy currents and so will be useful to very high frequencies.

The materials will be ferrimagnetic for all temperatures below a critical temperature, known as the *Néel temperature.* Above that temperature the materials also become *superparamagnetic*, similar to the way ferromagnetic materials behave above the Curie temperature. Since the two sublattices may have different temperature behavior, it is possible that at a given temperature the two moments may be equal but opposite in sign, as illustrated in Fig. 1.5. At this temperature, known as the *compensation temperature*, the two sublattices have equal magnetization so that the net magnetization is zero. This magnetization is the magnitude of the difference between the two sublattice magnetizations and will be positive, since above or below the compensation temperature, the material will become magnetized in the direction of an applied field.

The compensation temperature occurs above, below, or at room temperature, depending on the elements in the crystal. Unlike the demagnetized state above the Curie temperature, this state "remembers" its magnetic state, and changing its temperature from the compensation temperature reproduces the previous magnetic state. This property is useful in magneto-optical disks to render the stored information impervious to stray fields. This is done by choosing a compensation temperature that is close to the storage temperature. For practical devices the storage temperature is usually room temperature.

1.4 MICROMAGNETISM

In this section we assume that the temperature is fixed so that material parameters, such as saturation magnetization, may be regarded as constants. We then compute the equilibrium magnetization patterns in a ferromagnetic medium. The dynamics of magnetization are discussed in later sections. Thus, we choose the magnetization variation that minimizes the total energy. This total energy is the sum of the exchange energy, the magnetocrystalline anisotropy energy, and the Zeeman energy.

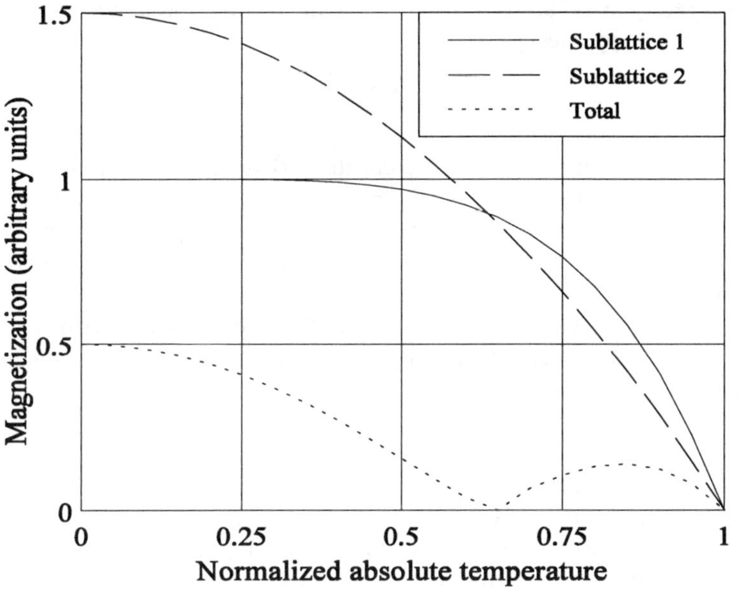

Figure 1.5 A ferrimagnetic material with a compensation temperature of approximately 65% of its Néel temperature.

The exchange energy, the source of the ferromagnetism, is given by

$$W_{ex} = \sum_{n.n.} J\mathbf{S}_i \cdot \mathbf{S}_j, \tag{1.17}$$

where n.n. denotes that the sum is carried out over all pairs of nearest neighbors, J is the exchange integral, and **S** is the spin vector. Since the wave functions are not isotropic, the exchange energy is not only a function of the difference in orientation of adjacent spins, but is also a function of the direction of the spins. Since a spin interacts with several nearest neighbors, the orientation energy depends upon the crystal structure. This variation in the exchange energy with spin orientation is called the *magnetocrystalline anisotropy energy*. We take it into account by adding an anisotropy energy density term to (1.17). For cubic crystals, the simplest form of this is given by

$$W_{cubic} = K(\alpha_x \alpha_y + \alpha_y \alpha_z + \alpha_z \alpha_x), \tag{1.18}$$

where the α's are the direction cosines with respect to the crystalline axes, and K is the anisotropy constant. If K is positive, the minimum anisotropy energy density occurs along each of the three axes of the crystal. On the other hand, if K is negative, the minimum anisotropy energy density occurs along the four axes that make equal angles with respect to the three crystal axes. Higher order terms may be added to this in certain cases.

Another type of anisotropy energy density that commonly occurs is the uniaxial anisotropy energy density. This is given by

$$W_u = K_u \sin^2\theta , \tag{1.19}$$

where K_u is the uniaxial anisotropy constant, θ is the angle the magnetization makes with respect to the z axis, and z is the easy axis if K_u is positive. If K_u is negative, then z is the hard axis, and the plane perpendicular to the z axis is the easy plane. We will denote the anisotropy energy density by W_{anis} whether it is cubic or uniaxial.

For the present, we will consider only one additional energy term, the *Zeeman energy*, which is the energy that a magnetic dipole **m** has due to a magnetic field. This energy is given by

$$W_{Zeeman} = -\mathbf{m}\cdot\mathbf{B}, \tag{1.20}$$

where **B** is the total magnetic field, which is the sum of the external applied field and the demagnetizing field of the body. We will decompose this term into the sum of the applied field energy and the demagnetizing energy. The energy of a magnetized body in an external field is given by

$$W_H = \int_V \mathbf{B}\cdot\mathbf{H}\, dV. \tag{1.21}$$

Since **B** is $\mu_0(\mathbf{H}+\mathbf{M})$, and since M^2 is constant, by choosing a different reference energy, this reduces to

$$W_H = \mu_0 \int_V \mathbf{M}\cdot\mathbf{H} dV, \tag{1.22}$$

where **H** is that applied field and V is the volume of the material. Similarly, the self-demagnetizing energy is given by

$$W_D = \frac{\mu_0}{2} \int_V \mathbf{M}\cdot\mathbf{H_D} dV, \tag{1.23}$$

where $\mathbf{H_D}$ is the demagnetizing field.

Thus, the total energy of the body is given by

$$W = W_{ex} + W_{anis} + W_D + W_H. \tag{1.24}$$

The magnetization pattern is then determined by adjusting the orientation of the magnetization at each point in the material to minimize the total energy. In principle we could find the orientation of the magnetization of each atom in the medium, but unless the object is very small, this would involve too many computations. Instead, in micromagnetism, we will define a continuous function whose value at each atomic site is the magnetization of that atom.

Micromagnetism is the study of magnetization patterns in a material at a level of resolution at which the discrete atomic structure is blended into a continuum, but the details are still visible. Thus, the orientation of the magnetization in the medium is obtained from a continuous function defined over the medium. Summations are replaced by integrations, and differences by derivatives. In particular, if **r** is the position of an atom and **a** is the relative position of a neighbor, the exchange energy density between them is given by

$$w_{\mathrm{ex}} = -\lim_{a\to 0}\frac{2J\mathbf{s}(\mathbf{r})\cdot\mathbf{s}(\mathbf{r}+\mathbf{a})}{a^3}. \tag{1.25}$$

Since the magnetization and the spin vector are in the same direction, we can replace **s** by $s\mathbf{M}/M_S$, where s is the magnitude of the spin vector and M_S is the magnitude of **M**. Then, if we expand **S(r+a)** in a Taylor series, we get

$$\mathbf{s}(\mathbf{r}+a\mathbf{1}_x) = \frac{s}{M_S}\left[\mathbf{M}(\mathbf{r}) + a\frac{\partial\mathbf{M}(\mathbf{r})}{\partial x} + \frac{a^2}{2}\frac{\partial^2\mathbf{M}(\mathbf{r})}{\partial x^2} + \cdots\right], \tag{1.26}$$

where a is the distance to the nearest neighbor atom in the x direction and $\mathbf{1_x}$ is a unit vector in the **x** direction. Then

$$\mathbf{s}(\mathbf{r})\cdot\mathbf{s}(\mathbf{r}+a\mathbf{1}_x) = \frac{s^2}{M_S^2}\left[1 + a\mathbf{M}(\mathbf{r})\cdot\frac{\partial\mathbf{M}(\mathbf{r})}{\partial x} + \frac{a^2}{2}\mathbf{M}(\mathbf{r})\cdot\frac{\partial^2\mathbf{M}(\mathbf{r})}{\partial x^2} + \cdots\right]. \tag{1.27}$$

The first term in the Taylor series is a constant and can be omitted by choosing a different energy reference. Since

$$\mathbf{M}\cdot\frac{\partial\mathbf{M}}{\partial x} = \frac{1}{2}\frac{\partial M^2}{\partial x}, \tag{1.28}$$

and since M^2 is a constant, the second term in (1.27) is zero. If we sum the terms in the y and z directions as well, then for a simple cubic crystal, the total exchange energy becomes

$$\begin{aligned}\sigma W_{ex} &= -\frac{A}{M_S^2}\int_V \mathbf{M}\cdot\left(\frac{\partial^2\mathbf{M}}{\partial x^2} + \frac{\partial^2\mathbf{M}}{\partial y^2} + \frac{\partial^2\mathbf{M}}{\partial z^2}\right)dV \\ &= -\frac{A}{M_S^2}\int_V \mathbf{M}\cdot\nabla^2\mathbf{M}\,dV,\end{aligned} \tag{1.29}$$

where

$$A = \frac{Js^2}{a}. \tag{1.30}$$

Because of the additional atoms in a unit cell, for a body-centered cubic lattice the exchange constant A is twice the value of a simple cubic lattice, and for a face-centered cubic lattice it is four times the value of a simple cubic lattice.

It is noted that (1.29) is approximate in two respects. First, the Taylor series is truncated. Thus, the change in magnetization between adjacent atoms is assumed to be small to allow the series to converge rapidly. This assumption is usually valid. The second approximation is more subtle in that we are approximating a discrete function by a continuous function. Since M^2 is constant, the second derivative of the magnetization diverges at the center of a vortex. Thus, (1.29) would calculate an infinite energy, although $J\mathbf{s}(\mathbf{r})\cdot\mathbf{s}(\mathbf{r}+a\mathbf{1}_\mathbf{x})$ remains finite at the center of the vortex.

The equilibrium magnetization in a medium is obtained by varying the direction of the magnetization so as to minimize the total energy. This can be done by directly minimizing the energy or by solving the Euler–Lagrange partial differential equation corresponding to this variational problem. The resulting magnetization pattern is referred to as the *micromagnetic solution*. This calculation must be performed numerically, except for a few cases, two of which are discussed in the next two sections. This introduces an additional discretization error that calculates a finite energy at the center of the vortex. This energy is incorrect unless the discretization distance is the same as the size of the magnetic unit cell.

If one is interested in the details of the magnetization change when the applied field changes, the dynamics of the process must be introduced. Two such effects — eddy currents, in materials with finite conductivity, and gyromagnetism — are discussed later.

1.5 DOMAINS AND DOMAIN WALLS

An equilibrium solution to the micromagnetic problem in an infinite medium is uniform magnetization along an easy axis. Then, both the exchange energy and the anisotropy energy are zero. Such a region of uniform magnetization is called a *domain*. In an infinite medium that is not uniformly magnetized, we will now see that the equilibrium solution is the division of the medium into many domains that are separated by domain walls that have essentially a finite thickness. Domain walls of many types are possible, but in this section we discuss only the two simplest types: the *Bloch wall* and the *Néel wall*. Furthermore, domain walls are classified by the difference in the orientations of the domains that they separate, expressed in degrees. For brevity, we limit ourselves to 180° walls.

We will consider a domain wall whose center is at $x = 0$, which divides a domain that is magnetized in the y direction as x goes to infinity and that is magnetized in the $-y$ direction as x goes to minus infinity. As one goes from one domain to the other, if the magnetization rotates about the x axis, it remains in the plane of the wall, and the wall is said to be a *Bloch wall*. On the other hand, if the magnetization rotates about the z axis, the wall is said to be a *Néel wall*.

1.5.1 Bloch Walls

Let us consider a Bloch wall that lies in the *yz* plane and that separates two domains: one magnetized in the *y* direction and the other magnetized in the $-y$ direction. If the domain magnetized in the *y* direction lies in the region of positive *x*, and the domain magnetized in the $-y$ direction lies in the region of negative *x*, then the magnetization can be written as

$$\mathbf{M}(x) = M_S \left\{\cos[\theta(x)]\mathbf{1}_y + \sin[\theta(x)]\mathbf{1}_z\right\}, \tag{1.31}$$

with the boundary conditions $\theta(-\infty) = 0$ and $\theta(\infty) = \pi$. That is, the magnetization is in the *z* direction for large negative values of *x* and in the $-z$ direction for large positive values of *x*. Differentiating twice with respect to *x*, we have

$$\frac{\partial^2 \mathbf{M}}{\partial x^2} = -\mathbf{M}\left(\frac{\partial\theta}{dx}\right)^2, \tag{1.32}$$

so that

$$\mathbf{M}\cdot\frac{\partial^2 \mathbf{M}}{\partial x^2} = -M_S^2\left(\frac{d\theta}{dx}\right)^2. \tag{1.33}$$

If there is no applied field, and since there is no demagnetizing field, the Zeeman energy is zero. Summing the remaining energies, the anisotropy energy and exchange energy, from (1.29), the energy in a domain wall per unit area is as follows:

$$W = \int_{-\infty}^{\infty}\left[A\left(\frac{\partial\theta}{\partial x}\right)^2 + g[\theta(x)]\right]dx, \tag{1.34}$$

where $g[\theta(x)]$ is the volume density anisotropy energy function.

We obtain the domain wall shape by finding the $\theta(x)$, which minimizes this integral subject to the constraints that $\theta(-\infty) = 0$ and $\theta(\infty) = \pi$. This minimum is found, using the calculus of variations, by solving the corresponding Lagrange differential equation corresponding to the minimization of this integral. In this case, this is given by

$$\frac{dg(\theta)}{d\theta} - 2A\left(\frac{d^2\theta}{dx^2}\right) = 0. \tag{1.35}$$

If we integrate this from 0 to θ, since $g(0)$ is zero and since $d\theta/dx|_{x=-\infty}$ is zero, we obtain

$$g(\theta) = 2A\int_0^{\infty}\frac{d^2\theta}{dx^2}d\theta = A\int_{-\infty}^{x}\frac{d}{dx}\left(\frac{d\theta}{dx}\right)^2 dx = A\left(\frac{d\theta}{dx}\right)^2, \tag{1.36}$$

or

$$x = \int_0^{\theta}\sqrt{\frac{A}{g(\theta)}}d\theta. \tag{1.37}$$

For crystals with uniaxial anisotropy, from (1.19),

$$g(\theta) = K_u\cos^2\theta. \tag{1.38}$$

Then

$$x = \sqrt{\frac{A}{K_u}}\int_0^{\theta}\frac{d\theta}{\sin\theta} = \frac{l_w}{\pi}\ln\left(\tan\frac{\theta}{2}\right), \tag{1.39}$$

where l_w is the classical wall width given by

$$l_w = \pi\sqrt{A/K_u}. \tag{1.40}$$

For iron, this is approximately 42 nm, or roughly 150 atoms wide. Solving for θ, one gets

$$\theta = \tan^{-1}\left(\exp\frac{\pi x}{l_w}\right) = \text{gd}\left(\frac{\pi x}{l_w}\right) - \frac{\pi}{2}, \tag{1.41}$$

where gd, defined by this equation, is called the *Gudermannian*. Figure 1.6 plots θ as a function of *x*. It is seen that more than half of the rotation in angle takes place between $\pm l_w$. In fact, in the equal angle approximation all the rotation takes place between $\pm l_w$. Since for many magnetic materials l_w is the order of 0.1 μm, the domain wall is very localized. Substituting (1.36) into (1.34), we see that the total energy density per unit wall area is given by

$$w = 2\int_{-\pi/2}^{\pi/2} g(\theta)\frac{dx}{d\theta}d\theta = 2\int_{-\pi/2}^{\pi.2}\sqrt{Ag(\theta)}\,d\theta. \tag{1.42}$$

Thus, for uniaxial materials, this becomes

$$w = 2\int_{-\pi/2}^{\pi/2}\sqrt{AK_u\sin^2\theta}\,d\theta = 4\sqrt{AK_u}. \tag{1.43}$$

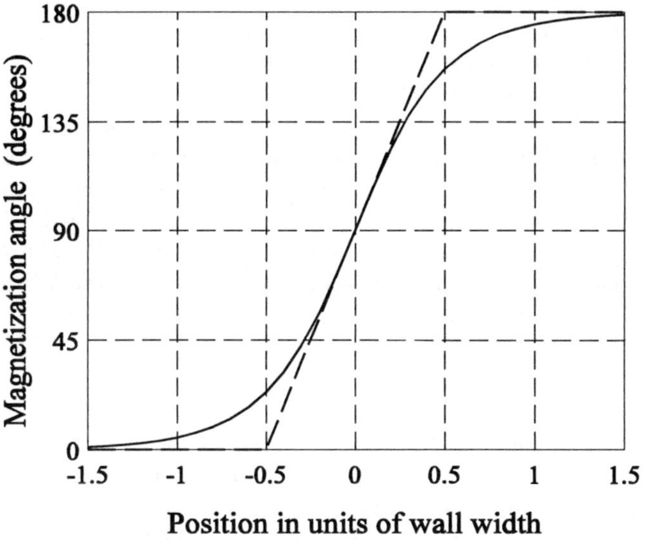

Figure 1.6 Variation of the magnetization angle for a Bloch wall: Dashed line indicates the equal angle approximation to the angle variation.

1.5.2 Néel Walls

For an infinite Néel wall, the magnetization is given by

$$\mathbf{M} - M_S\ [\cos\theta(x)\mathbf{1_x} + \sin\theta(x)\mathbf{1_z}], \tag{1.44}$$

where θ goes from 0 to π as x goes from $-\infty$ to ∞. The only difference between this and the Bloch wall is that the magnetization now turns so that when x is zero it points from one domain to the other. In this case, the divergence of **M** is no longer zero, and there is a Zeeman term in the total energy. Since the divergence of **B** is zero, the divergence of **M** is the negative of the divergence **H.** In particular,

$$\text{div } \mathbf{M} = \frac{\partial M_x}{\partial x} = M_S \cos\theta(x)\frac{d\theta}{dx} = -\text{div } \mathbf{H}. \tag{1.45}$$

Since **H** has only an x component, when we integrate this equation and use the boundary conditions that $\mathbf{H}(-\infty) = \mathbf{H}(\infty) = 0$, we are led to the conclusion that $H_x = -M_S$. From (1.23), the demagnetizing energy of the moments in this field is given by

$$W_D = -\frac{\mu_0 H_x M_S}{2} = \frac{\mu_0 M_x^2}{2} = \frac{\mu_0 M_S^2}{2}\cos^2\theta(x). \tag{1.46}$$

Comparison with (1.38) shows that this has the same variation as the uniaxial anisotropy energy. Thus, a Néel wall has the same shape as a Bloch wall whose anisotropy energy is given by $K_u + \mu M_S^2$. Since the wall energy is proportional to the square root of K_u, it is seen that the Néel wall will have greater energy than a Bloch wall. Thus, in infinite media, Bloch walls are energetically preferable to Néel walls. Furthermore, since the wall width is inversely proportional to the square-root of K_u, it is seen that the Néel wall will be thinner than a Bloch wall.

We have just discussed domain walls in infinite media. In finite media, the walls will interact with boundaries. Thus, in thin films, 180° walls between domains magnetized in the plane of the film tend to be Néel walls, to minimize demagnetizing fields. Furthermore, at the junction of two walls of opposite rotation, complex wall structures can form, such as cross-tie walls. This subject is beyond the scope of this chapter.

1.5.3 Coercivity of a Domain Wall

In the continuous micromagnetic case, the energy is not a function of the position of the domain wall. Thus the slightest applied field will raise the energy of the domain on one side of the wall with respect to the other, and there will be nothing to impede its motion, thus predicting zero coercivity. In a real crystal, the magnetization is not continuous because there are preferred positions of the domain wall, so there is a very small coercivity. The sources of coercivity in a real material are the imperfections in the crystal structure. We will briefly discuss imperfections of two types: *inclusions* and *dislocations* in the crystal lattice.

Inclusions are small "holes" in the medium, usually formed by the entrapment of bits of foreign matter. The inclusions either are nonmagnetic or have a much smaller magnetization than their surroundings. Such an inclusion will have magnetic poles induced on its surface, which will repel an approaching domain wall, thus impeding its progress. The equilibrium position of this wall in the absence of an applied field will be between the inclusions. The absence of exchange and anisotropy energy in the inclusion implies that the domain wall will have lower energy when it is situated on the inclusion also impeding its progress.

When a field is applied to a material with inclusions, the wall will bend in a direction that increases the volume of the domain that is closer to being parallel to the applied field. When the field is increased beyond a critical value, the domain will snap past that inclusion and become attached to another inclusion. We will denote the applied field behavior of the magnetization of the volume swept out by this motion as a *hysteron*. Even if it were possible to sweep that volume back, the field required to sweep the domain wall back generally would differ from the negative of the preceding field, which is now being restrained by different inclusions. Furthermore, these two fields are statistically independent of each other.

Dislocations in the crystal lattice also interact with domain walls. In some cases, the easy axes on the two sides of the dislocation may be aligned differently. This permits walls to be noninteger multiples of 90°. If the dislocations are sufficiently severe, the exchange interaction between atoms on the two sides of the wall may become negligible and a domain wall might not be able to cross the boundary.

A hysteron can switch either by rotation of the magnetization in the domain, as discussed in the next section, or by wall motion. In the latter case, if there is a wall, it has to be translated past the inclusions. On the other hand, if the material had been saturated, so that all the domain walls were annihilated, a new wall would have to be nucleated. The nucleation of a reversed domain requires a much higher field than that required to move a wall past each inclusion. Thus, nucleation usually takes place only when there are no domain walls anywhere in the crystal. If one measures the hysteresis loop of a material by controlling the rate of change of magnetization to a very slow rate, the field required for the initial change in magnetization is found to be larger than that needed for subsequent changes in magnetization. The resulting loop is said to be *reentrant*. Such a loop is shown in Fig. 1.7. The random variation in width is due to the variation in coercivity from inclusion to inclusion.

1.6 THE STONER–WOHLFARTH MODEL

A magnetic medium consisting of tiny particles can have a much higher coercivity than a continuous medium with inclusions. A model to analyze this case by means of an ellipsoidal particle was proposed by Stoner and Wohlfarth [7], who used a theorem, shown by Maxwell, that the demagnetizing field of a uniformly magnetized ellipsoid is also uniform. Thus, it is possible to have an object in which the applied field, the demagnetizing field, and the magnetization are all uniform. This model is called the *coherent magnetization model.* Other magnetization modes are possible if the material is large enough, but for bodies whose largest dimension is smaller than the width of a domain wall, only the uniform magnetization mode

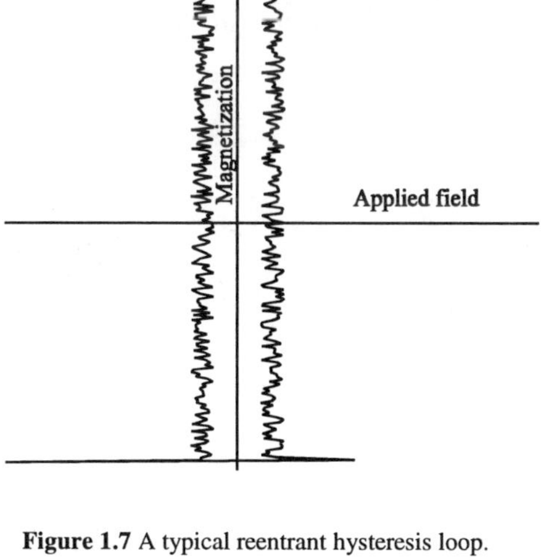

Figure 1.7 A typical reentrant hysteresis loop.

is possible. In such cases, we say that the particle is a *single domain particle*. Of course if the particle is too small, thermal energy might be sufficient to demagnetize it, and the particle would become *superparamagnetic*. That is, it would behave like a paramagnetic particle with a very large moment.

The Stoner–Wohlfarth model assumes that the particle is an ellipsoid and that its long (easy) axis is aligned with its magnetocrystalline uniaxial easy axis. It is also assumed that as the magnetization rotates, its magnitude remains constant. Because we assume that the particle is single domain, that is, it is uniformly magnetized, its exchange energy is seen to be zero. As the magnetization of the particle is rotated, the demagnetizing field changes in magnitude, and thus the demagnetizing energy changes because the demagnetizing factors along the different axes of the particle differ. This energy is referred to as *shape anisotropy energy*. Then magnetization will be oriented in such a way that the total energy — the sum of the applied field energy, the demagnetizing energy, and the shape anisotropy energy — is minimized. The sum of the latter two energies will be referred to simply as the *anisotropy energy*.

We will assume that a field is applied horizontally to a particle whose long axis makes an angle β with it, as shown in Fig. 1.8. All angles are measured in the counterclockwise direction, so that θ, the angle the magnetization makes with respect to the particle's long axis, as pictured, is negative. We will presently see that if the applied field is zero, the magnetization will lie along the easy axis of the particle; however, it could be oriented either way along that axis. Thus, the anisotropy energy will be doubly periodic as the magnetization rotates. We will also see that the applied field energy is unidirectional and thus is singly periodic.

Maxwell showed that for a uniformly magnetized general ellipsoid, the demagnetizing field is also uniform, though not antiparallel to it. The demagnetizing field can be written as the product of the demagnetization tensor and the magnetization. The demagnetization tensor is diagonalized if the coordinate axes are chosen to be the principal axes of the ellipsoid. In that case, the diagonal elements are referred to as the demagnetizing factors, and the demagnetizing field

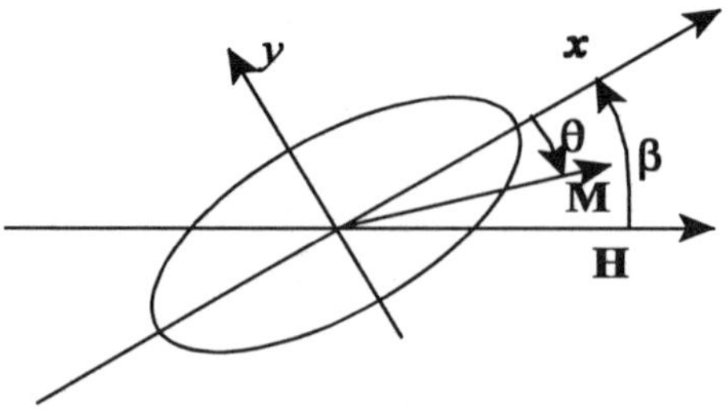

Figure 1.8 Stoner–Wohlfarth description of a spheroidal particle.

H_D is given by

$$H_D = D_x M_x \mathbf{1}_x + D_y M_y \mathbf{1}_y + D_z M_z \mathbf{1}_z, \tag{1.47}$$

where D_x, D_y, D_z are the demagnetizing factors along the three principal axes of the ellipsoid. Maxwell also showed that

$$D_x + D_y + D_z = 1. \tag{1.48}$$

For a spheroid, an ellipsoid of revolution, if the y and z are the two equal axes, then

$$D_y = D_z = \frac{1 - D_x}{2}. \tag{1.49}$$

It is well known that for this spheroid

$$D_x = \frac{1}{\alpha^2 - 1}\left[\frac{\alpha}{\sqrt{\alpha^2 - 1}} \ln\left(\alpha + \sqrt{\alpha^2 - 1}\right) - 1\right], \quad \text{for } \alpha > 1, \tag{1.50}$$

and

$$D_x = \frac{1}{1 - \alpha^2}\left[1 - \frac{\alpha}{\sqrt{1 - \alpha^2}} \sin^{-1}\sqrt{1 - \alpha^2}\right], \quad \text{for } \alpha < 1, \tag{1.51}$$

where α is the ratio of the length of the particle along the x axis to the length of the particle along other axes (see Bozorth [8]). It can be shown that as α approaches one for both formulas, the demagnetization factor approaches 1/3, the value for a sphere. It can also be shown that when $\alpha = 0$, then $D_x = 1$, and for large α (1.50) becomes

$$D_x = \frac{1}{\alpha^2}(\ln 2\alpha - 1), \tag{1.52}$$

and thus, goes to zero essentially as $1/\alpha^2$. A graph of D as a function of α, illustrating that $0 \le D_x \le 1$, is shown in Fig. 1.9.

Using the variables illustrated in Fig. 1.8 and the expression for demagnetizing energy in (1.23), it is seen that the demagnetizing energy is given by

$$W_D = \frac{\mu_0}{2}\mathbf{M}\cdot\mathbf{H_D}V = -\frac{\mu_0 M_S^2 V}{2}\left(D_x \cos^2\theta + \frac{1 - D_x}{2}\sin^2\theta\right). \tag{1.53}$$

If D_x is less than 1/3, then W_D is a minimum when $\theta = 0$. If the applied field is now nonzero, then we have to add an applied field energy, W_H, to this, where according to (1.21),

$$W_H = -\mu_0 \mathbf{H}\cdot\mathbf{M} = \mu_0 M_S [H_x \cos\theta + H_y \sin\theta]. \tag{1.54}$$

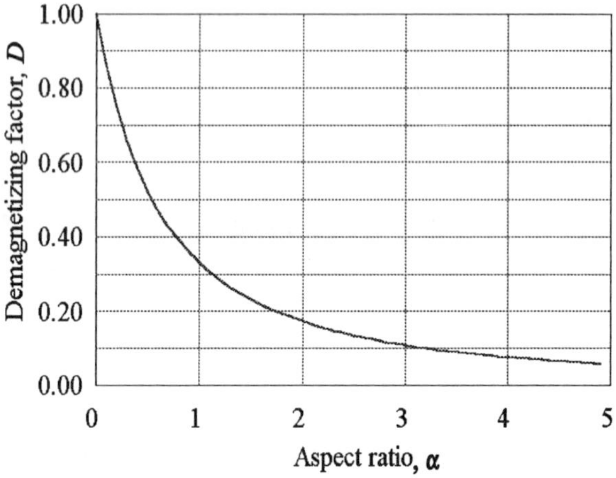

Figure 1.9 The demagnetizing factor of a spheroid as a function of its aspect ratio.

If the body remains uniformly magnetized, then the exchange energy is constant. Since the uniaxial magnetocrystalline anisotropy has the same spatial variation as the demagnetizing field, if their easy axes coincide, the two can be combined into a single term, and the effective demagnetizing factor must be increased by K_u. However, if the long particle axis does not line up with the magnetocrystalline axis, an effective easy axis between the two must be computed. A plot of the total energy, the sum of (1.53) and (1.54), is shown in Fig 1.10, for three applied field values: zero, $H_k/2$, and H_k, where $H_k = 2K/M$ is called the anisotropy field. It is seen that for zero applied field, the energy has two equal minima 180° apart. Then the magnetization could be oriented along either of these directions. As the field is increased, the minimum near 180° decreases in energy and moves to the left while the minimum near 0° increases and moves to the right. At the critical field, the minimum near 0° disappears, and above that field there is only a single minimum. When the field is decreased back to zero, the energy barrier between the two minima prevents the magnetization from going to the minimum near 0°. Thus, saturating a magnetic material is one method of putting it in a unique magnetic state.

In order to solve for the minimum energy, we take the total energy given by

$$W = -\mu_0 M_S V[H_x \cos\theta + H_y \sin\theta] - \frac{\mu_0 M_S^2 V}{2}\left[D_x \cos^2\theta + \frac{1-D_x}{2}\sin^2\theta\right], \quad (1.55)$$

differentiate it with respect to θ, and set it equal to zero. Thus, after dividing by $\mu_0 M_S V$, we get

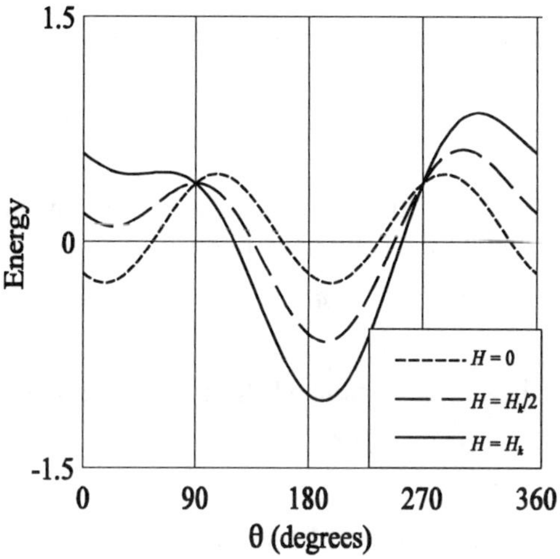

Figure 1.10 Energy as a function of magnetization angle for three applied fields.

$$\frac{1}{\mu_0 M_S V}\frac{\partial W}{\partial \theta} = H_x \sin\theta - H_y \cos\theta - C\sin\theta\cos\theta = 0, \tag{1.56}$$

where

$$C = M_S\left[\frac{1 - 3D_x}{2}\right]. \tag{1.57}$$

It is noted that for prolate particles, D_x is less than 1/3, so that C will be positive. To determine whether this is a minimum or a maximum, we take the second derivative of the energy with respect to θ, and obtain

$$\frac{1}{\mu_0 M_S V}\frac{\partial^2 W}{\partial \theta^2} = H_x \cos\theta + H_y \sin\theta + C(\sin^2\theta - \cos^2\theta). \tag{1.58}$$

Since the system seeks an energy minimum, this quantity must be positive at a stable equilibrium. To find the critical field, H_k, that is, the value of the field at which one of the minima disappears, we solve for the value that makes the second derivative zero. Thus, we obtain

$$H_x \cos\theta + H_y \sin\theta + C(\sin^2\theta - \cos^2\theta) = 0. \tag{1.59}$$

We can solve for $\cos\theta$ by multiplying (1.56) by $\sin\theta$, multiplying (1.59) by $\cos\theta$, and adding the results. Then one obtains

$$\cos\theta = (H_x/C)^{1/3} \quad \text{or} \quad H_x = C\cos^3\theta. \tag{1.60}$$

Similarly, we can solve for $\sin\theta$ by multiplying (1.59) by $\sin\theta$, multiplying (1.56) by $-\cos\theta$, and adding the results, yielding

$$\sin\theta = -(H_y/C)^{1/3} \quad \text{or} \quad H_y = -C\sin^3\theta. \tag{1.61}$$

Since $\sin^2\theta + \cos^2\theta = 1$, we can eliminate θ from (1.60) and (1.61). Thus,

$$H_x^{2/3} + H_y^{2/3} = C^{2/3}. \tag{1.62}$$

The solution to this equation is called the *Slonczewski asteroid* [9], which is illustrated in Fig. 1.11.

To determine the magnetization and its stability for a Stoner–Wohlfarth particle, one plots the vector magnetic field from the origin, as shown for two field vectors in Fig. 1.11. The direction of the magnetization is obtained by drawing a tangent from the asteroid to the tip of the field vector. The magnetization vector is obtained by drawing a vector whose length is given by $M_S V$ along that line. It is seen that when $\mathbf{H}_1$ is applied, the tip of the field vector falls outside the asteroid, and there is a unique state for the magnetization, indicated by $\mathbf{M}_1$; however, when $\mathbf{H}_2$ is applied, it falls inside the asteroid, and there are two stable states for the magnetization, both of which are indicated by $\mathbf{M}_2$.

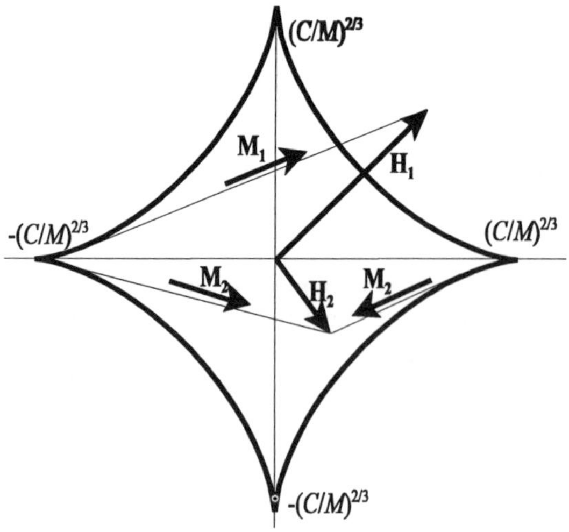

Figure 1.11 Slonczewski asteroid used to determine the state of a Stoner–Wohlfarth particle.

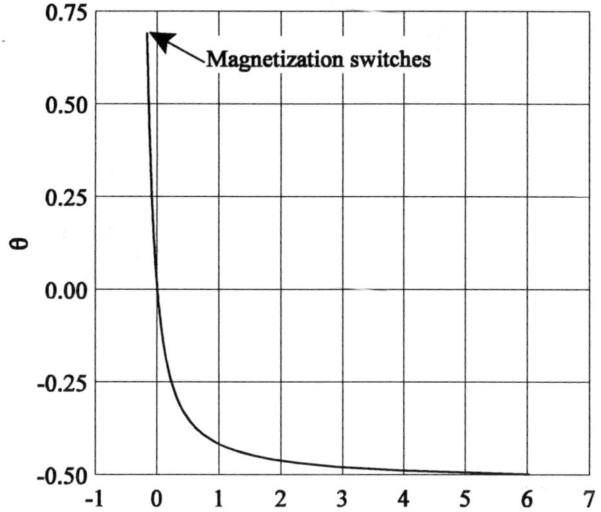

Figure 1.12 Variation of θ with the applied field for β = 0.5.

The normalized magnetization component along the applied field is given by

$$m = \cos(\theta + \beta). \tag{1.63}$$

The applied field that achieves this magnetization can be obtained by solving (1.56) as

$$H = -\frac{C\sin(2\theta)}{2\sin(\theta+\beta)}. \tag{1.64}$$

The variation of θ with applied field is illustrated in Fig. 1.12. It is seen that for positive fields, θ approaches $-\beta$ monotonically as the magnetization tries to align itself with the applied field. For negative fields, θ increases until it reaches its maximum, and then it switches.

We will define the critical angle θ_M as the angle at which the particle switches. It is obtained by solving for the value of θ that makes (1.58) equal to zero. It is thus possible to plot m as a function of H by varying θ between $-\beta$ and θ_M. That is, one must solve the transcendental equation

$$H\cos(\beta+\theta_M) - C\cos(2\theta_M) = 0. \tag{1.65}$$

If we substitute (1.64) into this, and use the tangent trigonometric identities, we obtain

$$\tan\theta_M = (\tan\beta)^{1/3}. \tag{1.66}$$

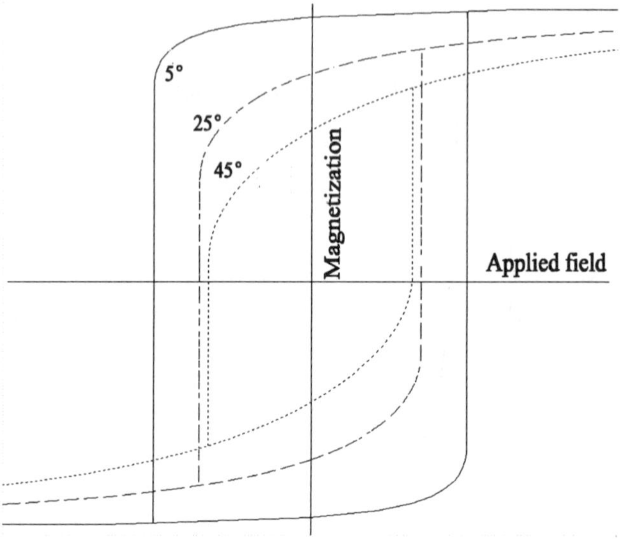

Figure 1.13 Possible Stoner–Wohlfarth particle hysteresis loops for $\beta = 2°$, $25°$, and $45°$.

If one plotted the component of the magnetization along the applied field's axis, that is, $M_S \cos(\theta + \beta)$, as a function of the applied field, one would obtain the hysteresis loops shown in Fig. 1.13 for three values of β. These loops show that for particles in the negative state, when the applied field reaches the critical field H_k, the particle abruptly switches to the positive state. If the magnetization was still negative before switching, this field is also the coercivity. On the other hand, if the magnetization was already positive, H_k is larger than that of the coercivity. The largest value of β for which H_k is equal to the coercivity is $45°$. It is seen that all the hysteresis loops have two critical fields that are the same in magnitude but opposite in sign.

The critical field of a particle as a function of particle angle β with respect to the applied field can then be computed, from (1.62), as

$$H_k = \frac{C}{(\cos\beta^{2/3} + \sin\beta^{2/3})^{3/2}}. \tag{1.67}$$

As shown in Fig. 1.14, this field is a maximum when $\beta = 0$ or $\pi/2$. When β increases from 0, the critical field of the particle decreases until $\beta = \pi/4$, and then increases back to the value it had at $\beta = 0$ when $\beta = \pi/2$.

For values of β beyond $\pi/4$, as the field is increased from negative saturation, the magnetization goes through zero before the magnetization switches. Thus, we have to distinguish between the critical field, the field at which the magnetization switches, and the coercivity, the field at which the magnetization is zero. The coercivity follows the critical field until $\pi/4$. Beyond that it obeys (1.64) with θ set

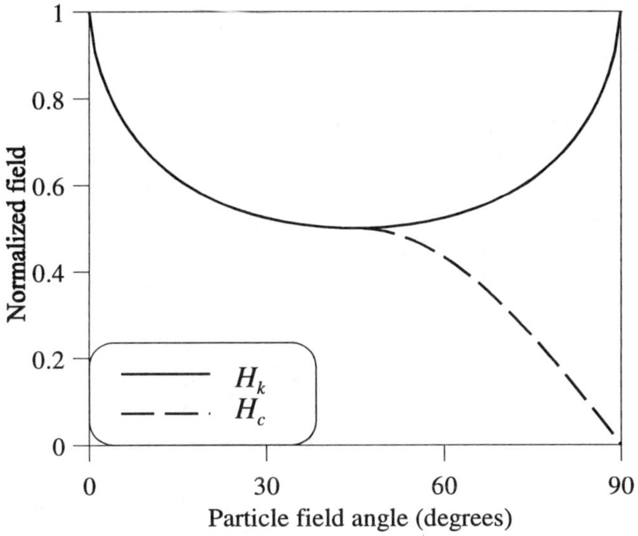

Figure 1.14 Coercivity and critical field variation with particle angle.

to the complement of β, as illustrated in Fig. 1.14. So the field at which the lower section of the curve crosses the H axis is a monotonic decreasing function of β.

For particles that are larger but still single domain, other nonuniform reversal modes are possible. These modes are characterized by smaller values of H_C and are sometimes referred to as *incoherent reversal modes*. Although these modes have a different β dependence, they have the same properties as the Stoner–Wohlfarth particles: two stable states, a monotonic decreasing function of H_C with β, and a maximum in H_C when β is 0 or π/2.

Real particles are generally ellipsoidal but with "corners." These corners permit magnetization reversals to be nucleated with fields considerably smaller than those necessary to nucleate reversals in ellipsoids. Since the shape of the particles prevents the existence of analytical solutions for them, reversal modes of these types have been studied numerically [10,11]. It was seen that for real particles, although their specific properties differ in magnitude and in various details, their general properties are the same as those of Stoner–Wohlfarth particles: that is, they have two stable states for a certain range of particle sizes; their switching field at first decreases with angle and then increases; and their coercivity is a monotonic decreasing function of angle.

One difference between ellipsoidal and nonellipsoidal particles is that for the latter there is *a nucleation volume* that, once reversed, causes the whole particle to reverse. This is also referred to as the *activation volume,* and it usually has an aspect ratio of unity. It may be thought of as the largest sphere that can be inscribed within the particle.

1.7 MAGNETIZATION DYNAMICS

Hysteresis is a rate-independent phenomenon; that is, the final state is the same no matter how fast the input changes to the final value. In fact, hysteresis is only a function of the field extrema. Thus, to obtain the possible final states, it is necessary only to solve the static equilibrium problem. To choose the particular magnetization pattern that is appropriate for a given input sequence if only hysteresis were involved, one would have to be sure only that the energy, in the sequence of magnetization patterns that were traversed by this magnetizing process, was a monotonically decreasing function of time. Other dynamic effects, which we will now discuss, may alter this sequence of equilibria.

There are two categories of dynamic effects: those that have time constants much slower than the rate of the applied field, and those that are comparable to or faster than the rate of the applied field. The former type includes magnetic aftereffect, which causes the magnetization to drift with time, while the latter type includes eddy currents and gyromagnetic effects. A rate-independent effect sometimes confused with these is accommodation. *Accommodation* is another process that causes the magnetization to drift; however, this process requires a change in applied field to trigger it. It is observed that repetitive minor loops apparently drift toward an equilibrium loop. As such, it is a rate-independent process and is discussed in Chapter 5.

Aftereffect refers to the slow change in magnetization with time that results from thermal processes. The magnetization is held in an equilibrium pattern by energy potential barriers. They may be surmounted by thermal energy according to the Arrhenius law. When this happens, the magnetization will find another local energy minimum. The higher the potential barrier, the longer it will take to be surmounted, but given enough time, any barrier may be surmounted. With this process, a magnetization pattern will change from a local energy minimum to a global energy minimum. For soft materials, with small energy barriers, this process will take the order of many minutes, but with harder materials, with correspondingly larger energy barriers, it may take centuries. This also is discussed in greater detail in Chapter 5.

1.7.1 Gyromagnetic Effects

We now turn our attention to gyromagnetic effects. When a magnetic field is applied to an electron, it creates a torque $\mathbf{T}$ on its magnetic moment $\mathbf{m}$ to align it with the magnetic field $\mathbf{B}$. That is,

$$\mathbf{T} = \mathbf{m} \times \mathbf{B} . \tag{1.68}$$

Since an electron also has an angular momentum, $\mathbf{k}$, we write

$$\mathbf{m} = -\frac{g\mu_0 e}{2m}\mathbf{k} = -\gamma\mathbf{k}, \tag{1.69}$$

where the minus sign is due to the sign of the charge of the electron, *e/m* is the ratio of the charge to the mass of an electron, and *g* is the gyromagnetic ratio, which is one for orbital motion and two for spin motion. The term γ is normally referred to as the *gyromagnetic ratio* of an electron. Thus, when an electron is subject to an applied magnetic field, its magnetization is unable to align itself with the field, but instead its magnetization precesses about the magnetic field. The precession frequency ω_0 is given by

$$\omega_0 = \gamma B. \tag{1.70}$$

This rotating magnetic moment radiates energy, thus permitting the electron to eventually align itself with the magnetic field. Therefore, the time rate of change of angular momentum is given by the Landau–Lifshitz equation

$$\frac{d\mathbf{k}}{dt} = -\gamma\mathbf{m}\times\mathbf{B} - \alpha\mathbf{m}\times(\mathbf{m}\times\mathbf{B}), \tag{1.71}$$

where α is the damping factor. For small damping factors, the moment will precess many times about the applied field, but for large damping factors, the moment will make a small fraction of a revolution about the applied field as it approaches equilibrium.

When an alternating rf magnetic field with frequency ω is applied to a material that is magnetized by a dc field acting along the *z*-direction, the material appears to have a nonreciprocal permeability tensor given by

$$[\mu] = \mu_0\begin{bmatrix} 1+\chi_{xx} & \chi_{xy} & 0 \\ -\chi_{xy} & 1+\chi_{xx} & 0 \\ 0 & 0 & 1 \end{bmatrix}, \tag{1.72}$$

where the reciprocal susceptibility is given by

$$\chi_{xx} = \frac{\omega_0\gamma B}{\omega_0^2 - \omega^2}, \tag{1.73}$$

and the nonreciprocal susceptibility is given by

$$\chi_{xy} = \frac{j\omega\gamma B}{\omega_0^2 - \omega^2}. \tag{1.74}$$

It is noted that B is the internal field in the material, which in ferromagnetic materials is given by

$$B = \mu_0 M D, \tag{1.75}$$

where D is the demagnetizing factor along the axis on which the material is magnetized. The nonreciprocal nature of this permeability permits one to build nonreciprocal passive devices, such as isolators, circulators, and other similar microwave devices.

1.7.2 Eddy Currents

When a field parallel to the magnetization on one side of a domain is applied, the domain wall experiences a "pressure" in a direction that would make the domain parallel to the applied field grow. In conductors, eddy currents are induced by Faraday's law whenever the applied field changes and consequently the magnetization changes. The eddy current field opposes the applied field and generally shields the interior of the material from it. For low frequencies, the applied field eventually penetrates the entire material. For high frequencies, the induced currents and the applied fields are limited to a very thin region close to the surface of the conductor, and so this effect is called the *skin effect*.

1.7.3 Wall Mobility

We will now address the question of how a domain wall moves in view of the constraints imposed by the Landau–Lifshitz equation. Consider a 180° Bloch wall between two domains magnetized in the $+z$ direction and the $-z$ direction. A z-directed field applies a pressure on the wall tending to move it in a direction such that the domain magnetized in the z direction would grow. This field would not apply a torque on the magnetic moments in either domain, since it has no component perpendicular to the magnetization. The atoms in the wall, however, experience a torque and will start to precess about the applied field. If this continues, the Bloch wall will become a Néel wall and will experience a demagnetizing field perpendicular to the applied field. The magnetic moments in the wall can now precess about this new field, and thus propagate the wall.

The larger the applied field, the faster the atoms in the domain wall will precess, and the more the Bloch wall will convert into a Néel wall. This will produce a larger demagnetizing field in the wall, causing it to precess faster, and thus the wall will move faster. Therefore, the wall's velocity will be proportional to the applied field, and its motion will be characterized by a mobility. This linear variation of wall velocity with the applied field terminates when the wall has completely converted to a Néel wall, and then the wall will have achieved a limiting velocity, referred to as the *Walker velocity*. This velocity depends on the material, but for most materials it is of the order of meters per second. The slowness of this motion was a limiting factor in bubble memories.

1.8 CONCLUSIONS

Modeling magnetic materials can be performed at various levels of detail: the atomic level, the micromagnetic level, the domain level, and finally at the nonlinear level. The first of these involves the use of quantum mechanics to compute the magnetization of individual electrons in atoms. The second level smears out the effect of individual atoms into a continuous function, and one can see the variation of the magnetization in the medium on a greater scale. At the domain level, the details of domain walls are invisible, and one sees only uniformly magnetized domains separated by domain walls of zero thickness. Finally, at the nonlinear level, one averages the magnetization over many thousands of atoms in order to replace the constituent equations that complete the definition of magnetic fields along with Maxwell's equation.

Preisach modeling, which we will describe in the subsequent chapters, falls into the nonlinear level of magnetization detail. This type of modeling describes not only gross effects, such as the major hysteresis loop, but also the details of minor loops. When coupled with the appropriate equations, it can describe dynamic effects as well. Finally, it can be coupled with phenomena of other types to describe hysteresis in such effects as magnetostriction.

The solution for the magnetization involves the calculation of the magnetic state of the system, since the behavior depends upon this. Then one can compute the magnetization of the system under the influence of an applied field when the magnetization is in this state. This type of problem is similar to a many-body problem, except that the system displays hysteresis. Thus, it can be referred to it as the *hysteretic many-body* problem.

In modeling coercivity, the quantities of interest are the discrete magnetization states and the Barkhausen jumps that occur when going from one state to another. The minimum change of state is the reversal of a single hysteron or magnetic entity. When there are many interacting hysterons, one is solving a hysteric many-body problem. Then one can go to the limit of a continuous density of hysterons. Preisach modeling is one of the mathematical tools for handling such densities.

The definition of the magnetic state will be based on the Preisach definition of *hysteron*, that is, a region that switches as a single entity and has two magnetic states. For hard materials, this region might be a single particle in particulate media or a single grain in thin-film media; for soft materials, it might be the volume switched by a single Barkhausen jump. A discrete entity with more than two states can be decomposed into several hysterons. Thus, the basic approach is identical for hard and soft materials, but the parameters chosen will differ. The classical Preisach model, which is discussed in the next chapter, is able to describe hysteresis in general, but the details do not accurately describe real-world phenomena. Subsequent chapters modify this model to correct these errors, using the physical principles just discussed.

References

[1] A. H. Morrish, *The Physical Principles of Magnetism*, Wiley: New York, 1965.

[2] S. Chikazumi with S. H. Charap, *Physics of Magnetism*, Wiley: New York, 1964.

[3] A. Visintin, *Differential Models of Hysteresis*, Springer-Verlag: Berlin, 1994.

[4] M. Brokate and J. Sprekels, *Hysteresis and Phase Transitions*, Springer-Verlag: New York, 1996.

[5] F. W. Sears, *Thermodynamics*, Addison-Wesley: Reading, MA, 1959.

[6] R. Becker and W. Döring, *Ferromagnetismus*, Springer-Verlag: Berlin, 1939.

[7] E. C. Stoner and E. P. Wohlfarth, "A mechanism of magnetic hysteresis in heterogeneous alloys," *Philos. Trans. R. Soc. London*, **A240**, 1948, pp. 599–642.

[8] R. M. Bozorth, *Ferromagnetism*, An IEEE Classic Reissue, IEEE Press: New York, 1994, p. 849.

[9] J. C. Slonczewski, *IBM Research Memorandum*. No. RM 003.111.224, October 1, 1956.

[10] M. E. Schabes and H. N. Bertram, "Magnetization processes in ferromagnetic cubes," *J. Appl. Phys.*, **64**, August 1988, pp. 1347–1357.

[11] Y. D. Yan and E. Della Torre, "Modeling of elongated fine ferromagnetic particles," *J. Appl. Phys.*, **66**, July 1989, pp. 320–327.

CHAPTER 2

THE PREISACH MODEL

2.1 INTRODUCTION

Hysteresis is a rate-independent branching nonlinearity; that is, the slope of the input–output curve depends only upon the sign of the rate of change of the input. Ferenc Preisach [1], developed a model [2] to explain hysteresis in soft magnetic materials. Although other models have been used, such as the play model discussed in Appendix A, they cannot give the physical insights into the magnetization process that are possible with this model. The Preisach model is capable of describing minor loops, as well as the major hysteresis loop; however, it is limited in its ability to describe magnetic materials by the *congruency property* and the *deletion property*. For this reason, many modifications of the Preisach model have been suggested. To differentiate between them, we will refer to the original Preisach model as the *classical Preisach model* and to the modifications of it as *Preisach-type models*.

2.2 MAGNETIZING PROCESSES

Before we begin the study of hysteresis in magnetic materials, we must define some of the common magnetizing processes needed to test the models we will develop. The first magnetizing process we will use, the *major hysteresis loop*, starts from a negative field large enough to saturate the material in the negative direction, goes to a positive field large enough to saturate the material in the positive direction, and then goes back. The section of the major hysteresis loop from negative saturation to positive saturation is called the *ascending major curve*. The other half of the major hysteresis loop is called the *descending major curve*. A typical major loop is illustrated by the solid line in Fig. 2.1.

The largest achievable magnetization is called the *saturation magnetization*, M_S. The magnetic field that increases the initial magnetization on the ascending major loop to zero is called the *coercivity*, H_C. The magnetization whenever the applied field is reduced to zero is called the *remanence*, M_{rem}. The *squareness* of the hysteresis loop, S, is the maximum remanence normalized to the saturation magnetization.

A hysteresis loop similar to the major loop is called the *remanence loop*. For a given applied field, a point on the remanence loop is measured by applying that field and reducing it to zero. The resulting remanence is plotted as a function of that applied field. Such a loop is illustrated by the dashed line in Fig. 2.1. The light dashed line indicates the relationship between the major loop and the remanent loop for a typical point. The magnetic field that increases the initial remanence to zero on the ascending remanent loop is called the *remanent coercivity*, H_{RC}. It is seen from the figure that H_{RC} is always larger than H_C.

The slope of the magnetization curve is called the *susceptibility* and denoted by χ. We will differentiate between the *remanent susceptibility* χ_r, the slope of the remanent curve, and χ. We will also define other susceptibilities later as needed.

If at some point on the ascending major loop the field is decreased, or if at some point on the descending major loop the field is increased, the locus of points on the magnetization field curve will enter the hysteresis loop. Such points are

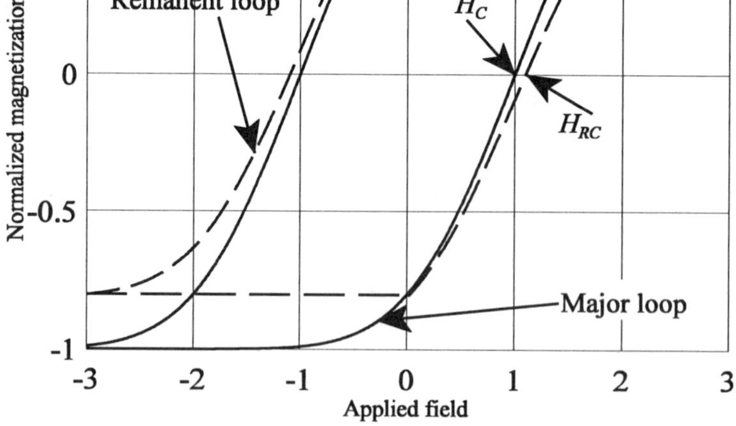

Figure 2.1 Major loop and remanent loop for material with unit coercivity.

called *turning points* and such traversals are called *first-order reversal curves*. A further reversal from one of these curves would be called a *second-order reversal curve*, and so on. A closed loop formed by two higher order reversal curves is called a *minor loop*.

A magnetization curve starting from the demagnetized state — that is, zero magnetization at zero field — and going to saturation is called a *magnetizing curve*. Such a curve is not unique but depends on how the material was demagnetized. We will reserve the name *virgin magnetizing curve* for the curve that starts from the state that was demagnetized by applying an *ac* field large enough to saturate the material and slowly reducing its magnitude to zero. This technique of obtaining a demagnetized state is called *ac demagnetization*.

2.3 PREISACH MODELING

The Preisach model considers the material to be a collection of square-loop hysterons, as shown in Fig. 2.2. The hysteron has a unique normalized magnetization, m, equal to one whenever the applied field H is greater than U, and a unique m equal to -1 whenever the applied field H is less than V. If the applied field lies between V and U, the magnetization may have either value depending on its history. Whenever the applied field increases beyond U, the magnetization state will switch to the positive state. Consequently the hysteresis loop is traversed, as shown by the arrows. Since the materials are passive, and since the energy loss is the area enclosed by the loop going in a counterclockwise direction, $U \geq V$ for all materials.

The *Preisach function* $P(U,V)$, where the *up-switching field* U and the *down-switching field* V are the coordinates defining the *Preisach plane*, is the density function of hysterons. With this definition, if a sufficiently large positive field is applied to the material, all the hysterons will be switched in the positive direction, and the resulting magnetization will be

$$\int\int_{U \geq V} P(U,V)\, dU\, dV = M_S. \tag{2.1}$$

If we define the Preisach function to be zero when $U < V$, we can integrate over the entire plane. For a negatively saturated material, the subsequent application of a positive field H_1 will switch all hysterons that have a U less than H_1. We will define the normalized Preisach function, $p(U,V)$, so that its integral is the normalized magnetization, that is, the magnetization divided by its saturation value. Then

$$\int_{-\infty}^{\infty} dU \int_{-\infty}^{U} dV\; p(U,V) = \int_{-\infty}^{\infty} dU \int_{-\infty}^{U} dV\; \frac{P(U,V)}{M_S} = 1. \tag{2.2}$$

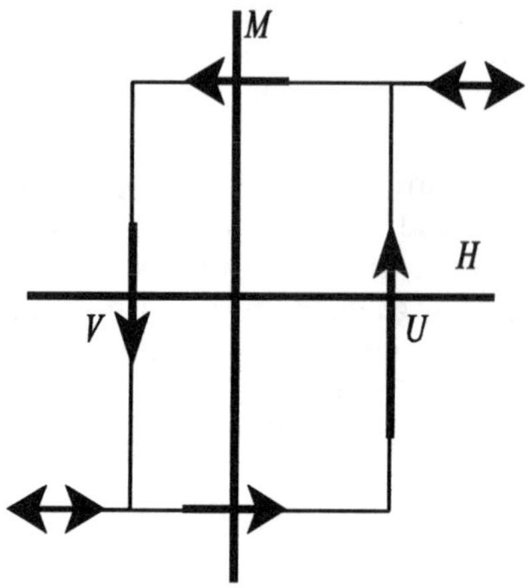

Figure 2.2 A typical hysteron in a Preisach model.

Let us consider a magnetizing process that starts from negative saturation followed by an applied field H_1. Then, since the change in magnetization when a hysteron switches from its negative value to its positive value is twice its magnitude, the normalized magnetization will be given by

$$m = \frac{M}{M_S} = -1 + 2 \int_{-\infty}^{H_1} dU \int_{-\infty}^{U} dV \; p(U,V). \tag{2.3}$$

The magnetization during field traversal to H_1 will follow the ascending part of the *major loop,* that is, the magnetization curve starting from negative saturation and going to positive saturation. If H_1 is not a saturating field, and the applied field is then decreased to a value H_2, the magnetization will follow a first-order reversal curve from H_1 to H_2. Subsequent traversals in the magnetization after additional reversals in the applied field are called *higher order reversal curves.*

Since the critical fields of an isolated hysteron, H_k and $-H_k$, must be the negative of each other, we say that to each of them is added an *interaction field* H_i to form U and V. Thus,

$$U = H_k + H_i \quad \text{and} \quad V = -H_k + H_i. \tag{2.4}$$

Since the interaction field varies as the magnetization of the other hysterons changes, one must be concerned with the stability of the Preisach function. More about this will be said in later chapters.

Starting from negative saturation, we will now obtain the sequence of magnetization due to the sequence of fields H_1, H_2, H_3, etc., as shown in Fig. 2.3. We note that the sequence of fields has the property that

$$H_k > H_{k+2} \text{ if } k \text{ is odd, and } H_k < H_{k+2} \text{ if } k \text{ is even}. \tag{2.5}$$

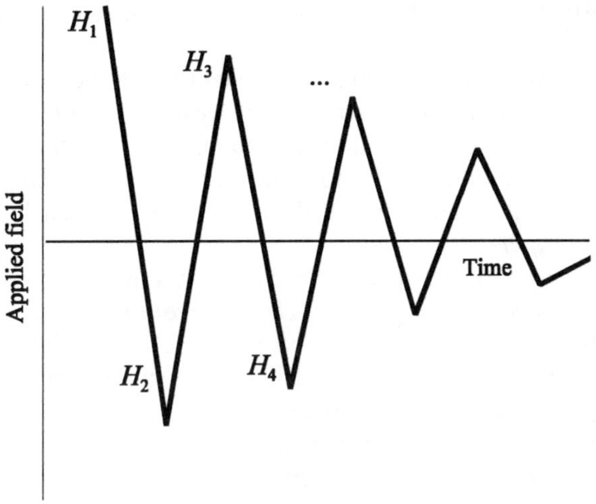

Figure 2.3 Arbitrary magnetizing process.

The normalized remanence after this sequence is applied is given by

$$m = \frac{M}{M_S} = -1 + 2 \int\int_{U<L} p(U,V)\, dV dU, \tag{2.6}$$

where L is the line illustrated in Fig. 2.4. The line is often described as the staircase dividing the Preisach plane between positively and negatively magnetized regions, and the corners on this line are referred to as the steps of the staircase. For odd k, if H_k were to be greater than H_{k-2}, the effect of H_k would be deleted. This property

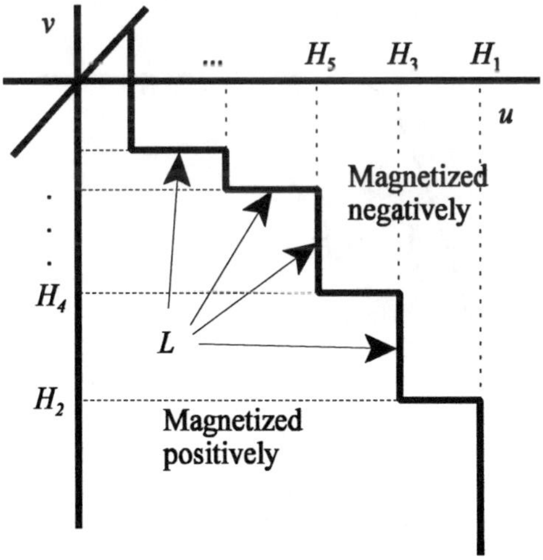

Figure 2.4 Division of the Preisach plane into a negatively magnetized region and a positively magnetized region.

of the Preisach model, known as the *deletion property*, is discussed in the next section. Similarly, the effect of a negative extremum is deleted by any subsequent more negative fields.

A *minor loop* is a magnetization curve that oscillates between two fields, H_1 and H_2. This curve may be obtained by any history prior to beginning this loop and so may be situated at any elevation inside the major loop. Three such loops are shown in Fig. 2.5. Section 2.7 will show that all these loops must be congruent to each other, if the process can be modeled by the Preisach model. This is known as the *congruency property*.

Mayergoyz has shown [3] that the congruency property and the deletion property are the *necessary and sufficient conditions* for a process to be representable by a classical Preisach model. Magnetic materials do not possess these properties, and to describe these processes accurately the Preisach model must be modified. This will be demonstrated later.

We now discuss several standard magnetizing processes that are referred to throughout this book. A *dc-magnetizing process* is the application of a *dc* magnetic field to a material and then its removal, leaving the material in a remanent state. The resulting remanence depends on the magnetic state of the material before the application of the field. If the material was saturated in the positive direction followed by a negative field, then this negative field is referred to as the *bias field*. Normally the bias field used is sufficiently negative to saturate the material in order to achieve a unique state, but other bias values can be used. The resulting remanence is computed from (2.6) where L is the line $U = H_1$.

An *anhysteretic magnetizing process* is one in which an *ac* and an offset *dc* magnetic field are simultaneously applied to the magnetic material as shown in Fig. 2.6. First the *ac* field is reduced to zero, and this is followed by the reduction of the

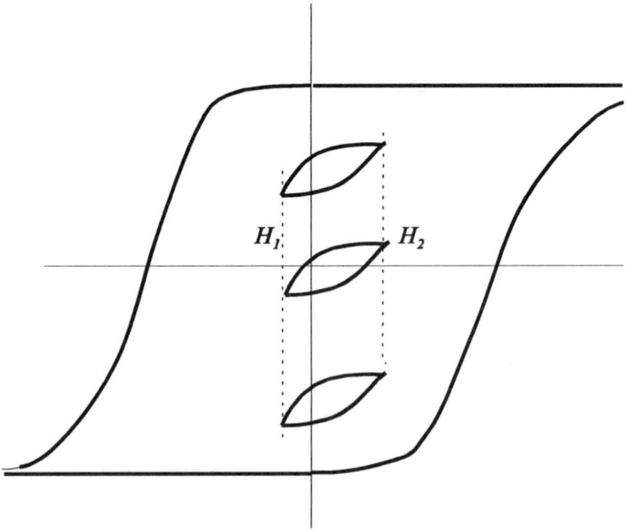

Figure 2.5 Minor loops oscillating between H_1 and H_2.

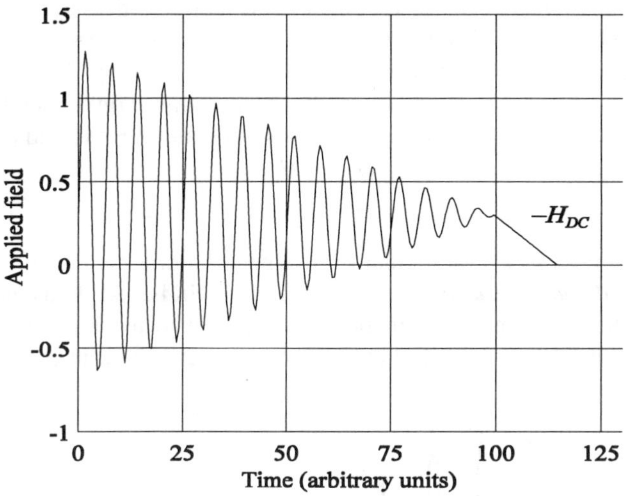

Figure 2.6 An anhysteretic magnetizing process as a function of time.

dc field to zero. We assume that the bias field is reduced so slowly that the applied field goes through many cycles as the *ac* field is reduced to zero.

Then the steps in the staircase on the Preisach diagram become very small and the staircase may be approximated by a straight line. If the *ac* field is large enough, then the resulting remanence is computed from (2.6), where the *dc* field is H_{dc}, and L is the line $U = -V + H_{dc}$ as illustrated in Fig. 2.7. The ellipse labeled "Preisach function" indicates a typical cross section of the Preisach function.

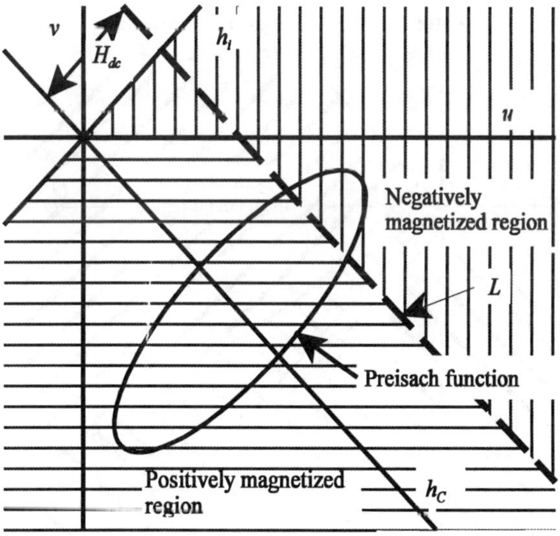

Figure 2.7 Division of the plane into positively and negatively magnetized regions by an anhysteretic magnetizing process.

An *ac-magnetizing process* is similar to the anhysteretic magnetizing process except that the two fields are simultaneously decreased to zero while the same proportion of their amplitudes is maintained, as shown in Fig. 2.8. In that case, if H_{ac} is the peak of the *ac* field and H_{dc} is the value of the *dc* field, then the resulting remanence is computed from (2.6), where L is the line $U = \rho V$, and ρ is given by

$$\rho = \frac{H_{dc} + H_{ac}}{H_{dc} - H_{ac}}, \tag{2.7}$$

as shown in Fig. 2.9. The particular *ac*-magnetizing process where H_{dc} is zero is known as *ac-demagnetization*, since the material will be left demagnetized if the Preisach function is an even function with respect to the line $U = -V$; that is, $P(U, V) = P(-V, -U)$. It is seen that if H_{ac} and H_{dc} are so large that the staircase

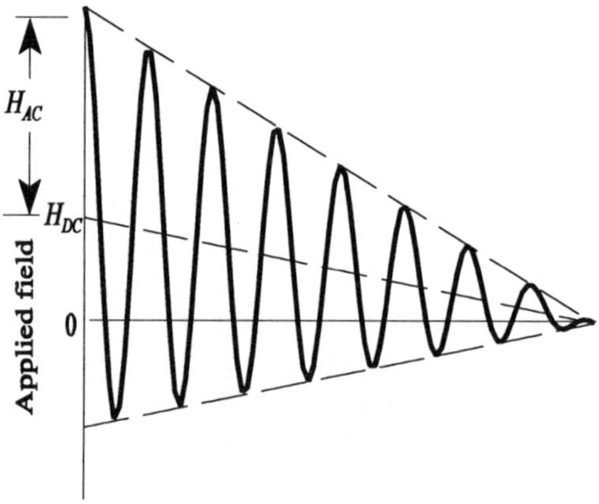

Figure 2.8 An *AC* magnetizing process.

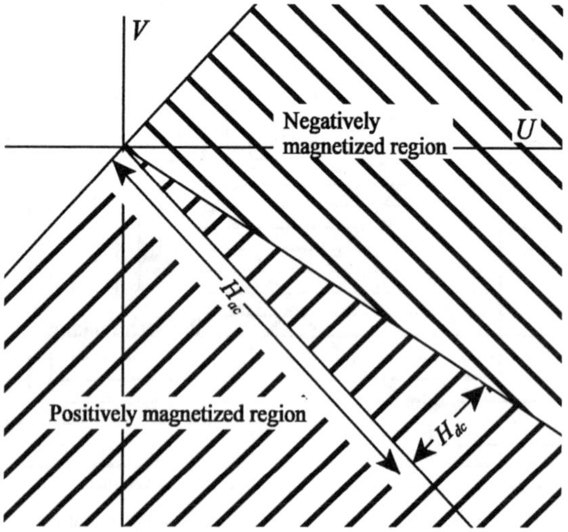

Figure 2.9 Division of the plane by the field in Fig. 2.8.

divides the entire nonzero portion of the Preisach plane, the resulting magnetization is only a function of the ratio of H_{ac} to H_{dc}.

A question sometimes raised is whether minor loops close on themselves. This is tested by the repetitive cycling between two applied fields, H_1 and H_2. Such a process is called an *applied field accommodation process.* If the minor loop thus traversed drifts with the cycle number, instead of closing on itself, the material is said to have accommodation. Other accommodation processes, also involving repetitive cycling, are defined later.

An important question in Preisach modeling is whether the Preisach function is stable at all: that is, whether the density of states is constant as the magnetization varies. It will be seen that although a constant Preisach function can be used to describe many observed magnetic hysteresis phenomena, such as finite anhysteretic susceptibility, it is not indeed constant. This instability in the Preisach function leads to violations of the congruency and deletion properties, as discussed in Chapters 4 and 5.

In some cases, the geometric interpretation of the Preisach model is cumbersome, so we now introduce the *Preisach state function Q* to facilitate our mathematical description. This function of the U and V is 1 if hysterons with these switching fields are magnetized in the positive direction and -1 if hysterons with these switching fields are magnetized in the negative direction. The state function can take intermediate values as well. For example, if a material is demagnetized by raising it above the Curie temperature, Q is zero everywhere. Later, we will also permit intermediate values of Q in the case of the accommodation and vector models. The net magnetization is given by

$$M = \iint_{U>V} Q(U,V)\,P(U,V)\,dU\,dV. \tag{2.8}$$

Using the normalized Preisach function, (2.2), we have

$$m = \frac{M}{M_S} = \iint_{U>V} Q(U,V)\,p(U,V)\,dU\,dV. \tag{2.9}$$

The state function Q will change during a magnetizing process, while P will not. In particular, if a material is saturated in the negative direction, then

$$Q(U,V) = -1 \tag{2.10}$$

for all points in the Preisach plane. In this case, the integral in (2.8) will become $-M_S$. If a positive field H is then applied, Q changes to

$$Q(U,V) = \operatorname{sgn}(H-U), \tag{2.11}$$

where the sign function sgn(x) is defined to be one if x is positive and minus one if x is negative. For the anhysteretic magnetizing process, when a large *ac* field in the presence of a *dc* field, H_{DC}, is reduced to zero, after which the *DC* field is reduced to zero, we have

$$Q(U,V) = \operatorname{sgn}(H_{dc} - U - V). \tag{2.12}$$

We will denote Preisach functions that are zero if U is negative or if V is positive as *single–quadrant Preisach functions*, since they are limited to the fourth quadrant of the Preisach plane. On the other hand, Preisach functions that extend into the first and third quadrants will be called *three-quadrant Preisach functions.* The hysteresis loops of single-quadrant Preisach functions have zero slope after a turning point. Three-quadrant functions do not decrease horizontally for $H > 0$, nor do they increase horizontally for $H < 0$. This behavior appears to be reversible but it is not, and therefore, it is referred to as *apparent reversible* behavior, as discussed further in Chapter 3.

2.4 THE PREISACH DIFFERENTIAL EQUATION

An alternative approach to computing the magnetization by integrating over the Preisach plane is the differential equation approach [4]. This method is very convenient for computing the magnetization as a function of time in real processes. The magnetic history of the material must be stored for the computation of the magnetization using the Preisach model. This can be done easily using a *push-down stack** for the extrema of the input. At the bottom of the stack is the largest magnitude applied field, and the successively smaller maxima and minima are stored above it until at the top of the stack is the current applied field. In the following analysis, we will assume that the applied field H is increasing and the stack contains the values H_1, H_2, H_3, etc. Since the magnetization changes by a factor of 2, in going from negative saturation to positive saturation, as long as H continues to increase and as long as $H < H_2$, the magnetization is computed as the solution to the following differential equation:

$$\frac{dm}{dH} = 2\int_{H_1}^{H} p(H,V)\,dV, \tag{2.13}$$

where H_1 is now the largest previous minimum. The upper limit could be set to infinity for physical Preisach functions, since $p(U,V)$ is zero whenever V is greater than U; however, if we use an artificial function for $p(U,V)$ that is nonzero when V is greater than U, we should leave the upper limit as H. Whenever $H = H_2$, we pop the top two values from the stack; that is, we set H_1 equal to H_3, H_2 equal to H_4, and so forth. The popping of the top two values from the stack is identical with the

*A stack is a programming tool in which data are stored in the order created rather than by position. A push-down stack is a last-in-first-out (LIFO) stack; that is, data are retrieved in inverse order from which they were stored. Data are said to be "pushed" on the stack when stored and "popped" from the stack when retrieved.

deletion property of the Preisach model. Then the process is computed by means of the same differential equation, but with a new lower limit on the integral.

If H starts to decrease, the present value of H is pushed on the stack; that is, we sequentially set H_1 equal to H, then H_2 equal to H_1, and so forth. Thus, H_1 is now the previous smallest undeleted maximum. Then, as long as H continues to decrease, and as long as $H > H_2$, the magnetization is computed as the solution to the similar differential equation

$$\frac{dm}{dH} = 2\int_{H}^{H_1} p(U,H)\ dU. \tag{2.14}$$

In this case, whenever $H = H_2$, we again pop the top two values from the stack, and continue.

If we are interested only in the normalized magnetization at the conclusion of a process, it can be expressed as an normalized Everett integral. In particular, if the process ends in $\cdots$, H_3, H_2, H_1, then the magnetization is given by

$$m(\cdots,H_3,H_2,H_1) = m(\cdots,H_3,H_2) + \mathrm{E}(H_2,H_1), \tag{2.15}$$

where $E(H_2,H_1)$ is the normalized Everett integral. If when changing the field from H_1 to H_2 no deletions of previous extrema occur, E is given by

$$\mathrm{E}(H_1,H_2) = 2\int_{H_1}^{H_2}\int_{H_1}^{U} p(U,V)\,dU\,dV. \tag{2.16}$$

It is noted that the sign of the integral is determined by whether H_l is larger or smaller than H_2.

2.4.1 Gaussian Preisach Function

A useful approximation for hard materials is to assume that the Preisach function is Gaussian, in both the interaction-free critical field H_k of the hysteron, and the interaction field H_i. Then this integral can be evaluated in closed form. The interaction field dependence can be justified on the basis of the central limit theorem of statistical theory, since the interaction field is the sum of the fields due to all the other hysterons, which are independent and identically distributed. The critical field dependence is an approximation to a log-normal dependence for the case when the mean critical field, $\bar{h}_k$, is more than twice its standard deviation, σ_k. The relationship between the Gaussian function and the log-normal function is discussed in Appendix B. Thus, we will assume that for hard materials the Preisach function is given by

$$p(H_k,H_i) = \frac{1}{2\pi\ \sigma_i\sigma_k}\exp\left\{-\frac{1}{2}\left[\frac{(H_k-\bar{h}_k)^2}{\sigma_k^2} + \frac{H_i^2}{\sigma_i^2}\right]\right\}, \tag{2.17}$$

where σ_k and σ_i are the standard deviations in the critical field and interaction field, respectively. We will later reserve lowercase h for operative fields, but since critical fields and operative critical fields are the same, we will use $\bar{h}_k$ for the average critical field to be consistent with later treatments. Since the critical fields and the interaction fields are independent phenomena, we expect their respective Preisach functions to have different means and standard deviations. Thus, the joint probability density will be the product of the individual density functions. It is noted that this function is valid to better than 0.5% if

$$\bar{h}_k > 2\sigma_k, \tag{2.18}$$

since the Preisach function must go to zero when H_k goes to zero. If this is not the case, one should use, for example, a log-normal function for the H_k variation. Alternatively, one can use a truncated Gaussian, but the normalization must be changed appropriately.

We can express this relationship in the U-V plane by using the inverse relationship of (2.4) between the U and V variables and the H_k and H_i variables:

$$H_k = \frac{U - V}{2} \quad \text{and} \quad H_i = \frac{U + V}{2}. \tag{2.19}$$

Noting that the Jacobian for the change in variables from H_k and H_i to U and V is 0.5, the Preisach function in terms of U and V is given by

$$p(U,V) = \frac{1}{4\pi\sigma_i\sigma_k} \exp\left[-\frac{\sigma_i^2(U-V-2\bar{h}_k)^2 + \sigma_k^2(U+V)^2}{8\sigma_i^2\sigma_k^2}\right]. \tag{2.20}$$

This may be rewritten

$$p(U,V) = \frac{1}{4\pi\sigma_i\sigma_k} \exp\left\{-\frac{(V+\lambda U+\kappa\bar{h}_k)^2}{2\tau^2} - \frac{(U-\bar{h}_k)^2}{2\sigma^2}\right\}, \tag{2.21}$$

where

$$\begin{aligned} \sigma &= \sqrt{\sigma_i^2 + \sigma_k^2}, \quad \tau = \frac{2\sigma_i\sigma_k}{\sigma}, \\ \lambda &= \frac{(\sigma_k^2 - \sigma_i^2)}{\sigma^2}, \quad \text{and} \quad \kappa = \frac{2\sigma_i^2}{\sigma^2} = 1 - \lambda. \end{aligned} \tag{2.22}$$

It is seen from (2.13) and (2.14) that the behavior of a hysteretic material depends on whether the applied field is increasing or decreasing. We will now compute the susceptibility for these cases separately.

2.4.2 Increasing Applied Field

When H is increasing, we carry out the V integration in (2.13) and set U equal to H to obtain the susceptibility χ, which is given by

$$\begin{aligned}\chi_{\text{inc}} = \frac{dm}{dH} &= \frac{1}{\sigma\sqrt{2\pi}}\exp\left[-\frac{(H-\bar{h}_k)^2}{2\sigma^2}\right]\left[\operatorname{erf}\left(\frac{V+\lambda H+\kappa\bar{h}_k}{\tau\sqrt{2}}\right)\right|_{V=H_1}^{V=H} \\ &= \frac{1}{\sigma\sqrt{2\pi}}\exp\left[-\frac{(H-\bar{h}_k)^2}{2\sigma^2}\right]\left[\operatorname{erf}\left(\frac{(1+\lambda)H+\kappa\bar{h}_k}{\tau\sqrt{2}}\right)\right. \\ &\qquad \left. -\operatorname{erf}\left(\frac{H_1+\lambda H+\kappa\bar{h}_k}{\tau\sqrt{2}}\right)\right],\end{aligned} \tag{2.23}$$

where the error function erf(x) is an odd, monotonically increasing function of x that approaches 1 as x approaches infinity, approaches -1 as x approaches minus infinity, and is defined by

$$\int_0^x \exp\left(-\frac{t^2}{2\tau^2}\right) dt = \tau\sqrt{\frac{\pi}{2}}\operatorname{erf}\left(\frac{x}{\tau\sqrt{2}}\right). \tag{2.24}$$

We note that at a reversal point, the upper limit is equal to the lower limit; thus, the susceptibility is zero. This property is true for any Preisach function.

If (2.18) holds, then the upper limit in (2.23) can be replaced by infinity, and χ_{inc} is given by

$$\chi_{\text{inc}} = \frac{dm}{dH}, \tag{2.25}$$

or

$$\chi_{\text{inc}} = \frac{1}{\sigma\sqrt{2\pi}}\exp\left[-\frac{(H-\bar{h}_k)^2}{2\sigma^2}\right]\left[1-\operatorname{erf}\left(\frac{H_1+\lambda H+\kappa\bar{h}_k}{\tau\sqrt{(2)}}\right)\right]. \tag{2.26}$$

The error in this is a function of how much larger $\bar{h}_k$ is compared to σ_k.

For the ascending major loop, H_1 is negative infinity, and the error function is -1. Thus, in this case, the major loop is an error function and its slope is a Gaussian. Then the major loop susceptibility is described by

$$\frac{dm}{dH} = \frac{1}{\sigma}\sqrt{\frac{2}{\pi}}\exp\left[-\frac{(H-\bar{h}_k)^2}{2\sigma^2}\right]. \tag{2.27}$$

Therefore, the ascending curve of the major loop is given by

$$m = \operatorname{erf}\left(\frac{H - \bar{h}_k}{\sigma\sqrt{2}}\right). \tag{2.28}$$

If we are traversing a minor loop, then at corners of the staircase, although m is continuous, dm/dH is not.

2.4.3 Decreasing Applied Field

When H is decreasing, we may rewrite (2.20) as

$$p(U,V) = \frac{1}{4\pi\sigma_i\sigma_k} \exp\left\{-\frac{(U+\lambda V-\kappa\bar{h}_k)^2}{2\tau^2} - \frac{(V+\bar{h}_k)^2}{2\sigma^2}\right\}, \tag{2.29}$$

and we can use (2.14) to rewrite (2.23) as

$$\chi_{\text{dec}} = \frac{dm}{dH} = \frac{1}{\sigma\sqrt{2\pi}} \exp\left[-\frac{(H+\bar{h}_k)^2}{2\sigma^2}\right]\left[\operatorname{erf}\left(\frac{U+\lambda H-\kappa\bar{h}_k}{\tau\sqrt{2}}\right)\right|_{U=H}^{U=H_1} \tag{2.30}$$

or

$$\chi_{\text{dec}} = \frac{1}{\sigma\sqrt{2\pi}}\exp\left[-\frac{(H+\bar{h}_k)^2}{2\sigma^2}\right]\left[\operatorname{erf}\left(\frac{H_1+\lambda H-\kappa\bar{h}_k}{\tau\sqrt{2}}\right) - \operatorname{erf}\left(\frac{(1+\lambda)H-\kappa\bar{h}_k}{\tau\sqrt{2}}\right)\right]. \tag{2.31}$$

If (2.18) holds, the lower limit can be replaced by minus infinity so that the second error function is -1, and then (2.31) can be approximated by

$$\chi_{\text{dec}} \approx \frac{1}{\sigma\sqrt{2\pi}} \exp\left[-\frac{(H+\bar{h}_k)^2}{2\sigma^2}\right]\left[1 + \operatorname{erf}\left(\frac{H_1+\lambda H-\kappa\bar{h}_k}{\tau\sqrt{2}}\right)\right]. \tag{2.32}$$

We conclude this section by computing three special cases to illustrate the dependence of first-order reversal curves, starting at $-\bar{h}_k$ from the descending major loop, on the standard deviations. In the first case σ_i is equal to σ_k, in the next σ_k is equal to zero, and in the last σ_i is equal to zero. In all the cases, we will assume that the value of σ is the same, but we will vary the ratio of σ_i to σ_k.

Case I: First we set σ_i equal to σ_k. Then, λ is zero and κ is 1. Thus, for increasing H, (2.26) becomes

$$\frac{dm}{dH} = \frac{1}{\sigma\sqrt{2\pi}} \exp\left[-\frac{(H-\bar{h}_k)^2}{2\sigma^2}\right]\left[1 - \operatorname{erf}\left(\frac{H_1+\bar{h}_k}{\tau\sqrt{2}}\right)\right]. \tag{2.33}$$

Case II: In the second case, if we set σ_k equal to zero, then, τ is equal to zero, κ is 2 and λ is -1. Since the argument of the error function now is

$(H_1 - H + 2\bar{h}_k)/\tau\sqrt{2}$, whose magnitude is infinite, the value is either $+1$ or -1 depending on the sign of the argument. When the error function is positive, the quantity in the square bracket is 2, but otherwise it is zero. Then

$$\frac{dm}{dH} = \begin{cases} \frac{1}{\sigma\sqrt{2\pi}} \exp\left[-\frac{(H-\bar{h}_k)^2}{2\sigma^2}\right] & \text{if} \quad H < H_1 + 2\,\bar{h}_k \\ 0 & \text{if} \quad H \geq H_1 + 2\,\bar{h}_k. \end{cases} \tag{2.34}$$

Case III: Finally, if we set σ_i equal to zero, then τ is again zero, but this time κ is zero and λ is 1. Since the argument of the error function is now $(H_1 - H)/\tau\sqrt{2}$, the magnitude of the error function is unity, and the sign in front of the error function is the same as the sign of its argument. Thus,

$$\frac{dm}{dH} = \begin{cases} \frac{1}{\sigma\sqrt{2\pi}} \exp\left[-\frac{(H-\bar{h}_k)^2}{2\sigma^2}\right] & \text{if} \quad H > -H_1 \\ 0 & \text{if} \quad H < -H_1. \end{cases} \tag{2.35}$$

This is the case for no interaction. According to Wohlfarth [5] the slope for the virgin curve, that is, the magnetization curve starting from the demagnetized state, should be half the slope of the major loop for the same field. In this case, it does not matter how the material was demagnetized.

A plot of these three special cases is shown in Fig. 2.10 for the case of a first-order reversal curve starting from $-\bar{h}_k$ on the descending major loop, using a value for σ of 0.35. It is seen that if $\sigma_k = 0$, then m remains zero until it meets the major loop. If $\sigma_i = 0$, there is no interaction, and the slope is half that of the major loop, as suggested by Wohlfarth. If $\sigma_k = \sigma_i$ the magnetization is half at $H = \bar{h}_k$. All these loops will then follow the major loop when they eventually encounter it. Finally, minor loops between the same extrema will be congruent, since the Everett integrals will not be a function of the magnetization.

It is noted from (2.26) that the slope at any given applied field depends upon the choice of H_1. For example, a specimen can be demagnetized in various ways. For an increasing field, if the specimen had been *ac* demagnetized, then H_1 is equal to H. If it had been *dc* demagnetized, the process would start from saturation, go to the opposite coercive field, and then go to zero. Then H_1 would be either $\bar{h}_k$ or ∞, depending upon whether the process had started from positive saturation or negative saturation. For other processes, other slopes are possible. The range of slopes is determined by σ_i, and the range is zero if σ_i is zero.

For the anhysteretic magnetizing process, it can be shown that the magnetization is given by

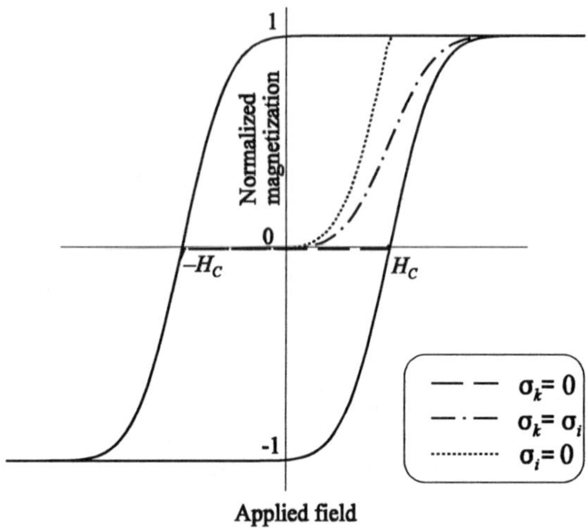

Figure 2.10 First-order reversal curves that originate from the descending major loop at the coercive field. Curves are shown for three pairs of values of σ_i and σ_k, but with the same σ.

$$m = \operatorname{erf}\left(\frac{H_{dc}}{\sigma_i\sqrt{2}}\right), \tag{2.36}$$

where H_{dc} is the previously defined offset field.

2.5 MODEL IDENTIFICATION: INTERPOLATION

To characterize a material by the Preisach model, one must first identify the Preisach function. If the type of function is unknown, the only recourse is to explore the entire Preisach plane. To be able to use the Preisach model to compute the magnetization, first one must know the saturation magnetization. This can be accomplished easily by measuring the magnetization in a large field. Then the Preisach function can be normalized by (2.2), so that $P(U,V)$ is equal to $M_S p(U,V)$. The identification is then performed by using first-order reversal processes; that is, for various H_1 and H_2, one starts from negative saturation, then applies a field H_1, followed by H_2, which is less than H_1. This magnetization is given by $M(H_1, H_2)$. Thus, for small ϵ's, we have

$$p(H_1,H_2) = \frac{M(H_1+\epsilon_1,H_2+\epsilon_2)+M(H_1,H_2)-M(H_1+\epsilon_1,H_2)-M(H_1,H_2+\epsilon_2)}{\epsilon_1\epsilon_2}. \tag{2.37}$$

In the limit as ϵ goes to zero, this becomes

$$p(U,V) = \frac{\partial^2 M(U,V)}{\partial U \, \partial V} . \tag{2.38}$$

Alternately, we can express the Preisach function in terms of Everett integrals; that is,

$$p(H_1,H_2) = \lim_{\epsilon \to 0} \frac{\mathrm{E}(H_2+\epsilon,H_1) - \mathrm{E}(H_2,H_1)}{\epsilon} = \frac{\partial \mathrm{E}(H_2,H_1)}{\partial H_2} . \tag{2.39}$$

An alternate method for computing the Preisach function [6] utilizes its symmetry. Since there is no preferred direction of magnetization, for a classical Preisach model, we must have

$$p(u,v) = p(-v,-u); \tag{2.40}$$

that is, the Preisach function must be symmetrical about the $u = -v$ axis. Consider an *ac*-demagnetized sample that is then subject to an anhysteretic magnetizing process, starting from the point $U = H_1$ and $V = H_2$. We will denote the normalized remanence at the conclusion of this process by $m_{\mathrm{anhys}}(H_1,H_2)$. In a fashion similar to the derivation of (2.38), it can be shown that

$$p(U,V) = \frac{\partial^2 m_{\mathrm{anhys}}(U,V)}{\partial U \, \partial V} . \tag{2.41}$$

The problem with both these approaches is that experimentally one has to take second differences of measured values. The error in taking second differences is much larger than the error in making the measurements. Thus, in the next section we introduce a method that is much less sensitive to errors.

2.6 MODEL IDENTIFICATION: CURVE FITTING

The preceding method of identifying the Preisach function required taking the second partial derivative of the magnetization resulting from a first-order reversal curve. This method is very prone to experimental errors. Furthermore, if one wants to obtain the Preisach function for the entire plane, one has to map out the entire plane. An alternate method of identifying the Preisach function is to assume that, similar to (2.17), it is of the form

$$P(H_k,H_i) = \frac{M_S}{2\pi \; \sigma_i \sigma_k} \exp\left\{ -\frac{1}{2}\left[\frac{(H_k - \bar{h}_k)^2}{\sigma_k^2} + \frac{H_i^2}{\sigma_i^2} \right] \right\} . \tag{2.42}$$

This function has four unknown parameters: M_S, $\bar{h}_k$, σ_k, and σ_i. If we can determine these parameters directly, then we know the Preisach function over the entire plane.

The first two parameters can be obtained from the major loop: M_S is the asymptotic value of the magnetization for large fields, and $\bar{h}_k$ is the value of the applied field that reduces the magnetization to zero. The other two parameters, σ_k, and σ_i, must be obtained in two steps: first, σ^2, the sum of their squares, is obtained by fitting the major loop, and then their ratio is obtained by measuring a first-order reversal curve. The first step is performed by fitting a Gaussian curve to the derivative of the major loop. The mean of this Gaussian is another measure of $\bar{h}_k$ and its standard deviation is a measure of σ.

To separate σ into its two parts, let us measure the magnetization at the conclusion of the process that starts from positive saturation, reduces the field to $-\bar{h}_k$, and then follows the first-order transition back to $\bar{h}_k$ [7]. At the conclusion of this process, m is equal to $E(-\bar{h}_k, \bar{h}_k)$. Thus,

$$m = E(-\bar{h}_k,\bar{h}_k) = \int_{-\infty}^{\bar{h}_k} dU \int_{-\bar{h}_k}^{\infty} dV\, p(U,V) = \int_{-\infty}^{\bar{h}_k'} dH_k \int_{\bar{h}_k-H_k}^{-(\bar{h}_k-H_k)} dH_i\, p(H_k,H_i). \quad (2.43)$$

Let us define

$$\rho = \sigma_k/\sigma_i. \quad (2.44)$$

Substituting this and (2.22), into (2.17) gives us the following expression for the Preisach function:

$$p(H_k,H_i) = \frac{2}{\pi\sigma\tau}\exp\left[-\frac{(H_k-\bar{h}_k)^2+\rho^2 H_i^2}{2\sigma^2\tau^2}\right]. \quad (2.45)$$

If we make the substitution using the dummy variables r and θ, where

$$H_k-\bar{h}_k = r\cos\theta \quad \text{and} \quad H_i = r\sin\theta, \quad (2.46)$$

we obtain

$$\begin{aligned} m &= \frac{2}{\pi\sigma\tau}\int_0^{\infty} dr \int_0^{\pi/4} d\theta\, r\exp\left[-\frac{r^2(\cos^2\theta+\rho^2\sin^2\theta)}{2\sigma^2\tau^2}\right] \\ &= \frac{2\rho}{\pi}\int_0^{\pi/4}\frac{d\theta}{\cos^2\theta+\rho^2\sin^2\theta} = \frac{2}{\pi}\tan^{-1}\rho. \end{aligned} \quad (2.47)$$

Thus,

$$\rho = \frac{\sigma_k}{\sigma_i} = \tan\left(\frac{m\pi}{2}\right), \quad \sigma_i = \sigma\cos\left(\frac{m\pi}{2}\right), \text{ and } \sigma_k = \sigma\sin\left(\frac{m\pi}{2}\right). \quad (2.48)$$

Since m varies between zero and one, both σ_i and σ_k vary between zero and σ. For the three cases shown in Fig. 2.10, m at $\overline{h_k}$ has this property. This identification method does not use any differentiation to obtain the Preisach function and furthermore can integrate many observations to obtain the parameters, further improving accuracy.

2.7 THE CONGRUENCY AND THE DELETION PROPERTIES

We now show that the congruency property and the deletion property are the necessary and sufficient conditions for a process to be modeled by a Preisach model, as was first shown by Mayergoyz [3]. As stated earlier, one property of the classical Preisach model is that all minor loops between the same pair of applied fields are congruent. From Fig. 2.11, it can be seen that cycling between the two applied fields H_1 and H_2 divides the Preisach plane into four regions. The region R_2 is always set negative by H_2, and the region R_3 is always set positive by H_1, while the region R_1 alternates between positive and negative as the applied field is cycled. Region R_4 on the other hand is unaffected by this process. This latter region determines only the position of the minor loop within the major loop. Thus, the congruency property is a necessary condition for a process to be described by a Preisach model. A consequence of this analysis is that minor loops are always contained within the major loop.

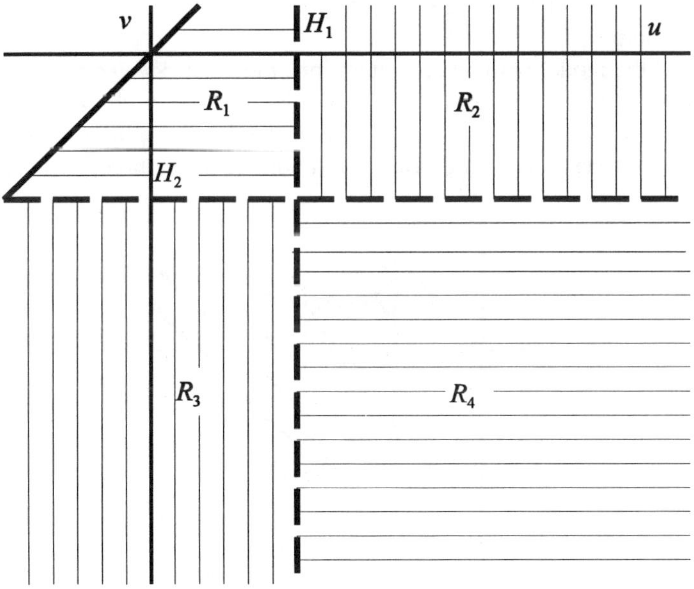

Figure 2.11 Division of the plane to illustrate the congruency property.

The deletion property can be understood by means of the process illustrated in Fig. 2.12. In this case we have a staircase line dividing the Preisach plane into two regions: the union of R_1 and R_3, a positively magnetized region, and the union of R_2 and R_4, a negatively magnetized region. It is assumed that the smallest positive corner of the staircase is at H_2. When a field H_1 is applied, which is less than H_2, the region R_3 is then switched from negative to positive, but the corner at H_2 still maintains its identity. When the applied field is increased to H_3, the region R_4 is now switched, so that it becomes part of R_I thereby deleting the effect of H_2. Thus, we have illustrated that the deletion property is a necessary condition for a process to be described by a Preisach model.

To show that these are also the sufficient conditions, we will show that they uniquely determine the Preisach function. That is, a process that possesses the congruency and deletion properties is capable of defining a unique Preisach function. This can be seen because the deletion property ensures that the saturation state is unique, and the congruency property ensures that the Everett functions are unique. Thus, the Preisach function determined by (2.39) is unique. We have now shown that these two properties are the necessary and sufficient conditions for a process to be described by a classical Preisach model.

In Chapters 4 and 5 we discuss the absence of both congruency and deletion properties in the magnetic properties of real materials. This does not mean that we cannot use the Preisach framework for describing them; but instead, we show that the Preisach model can be used as an element in the description of the entire process. The alterations that we will make to the Preisach model will be based on physical principles.

We call the model without any alterations the *classical Preisach model*. The classical model has some additional properties that are a characteristic of the unaltered model only. First of all, because of the deletion property, minor loops retrace themselves after the first iteration. Furthermore, whenever the

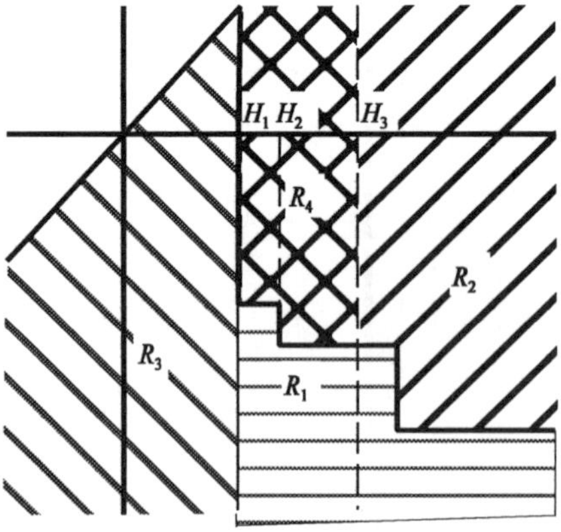

Figure 2.12 Division of the Preisach plane that illustrates the deletion property.

magnetization changes direction, the susceptibility instantly goes to zero and then increases again. The ascending major loop is continuous and has a continuous first derivative. For single-quadrant media, the magnetization is constant until the field reaches zero, but for three-quadrant media, the magnetization starts changing sooner and has a finite slope at zero field. Also, the small-signal susceptibility is only a function of the applied field. All these limitations are violated to some extent in real media, and these limitations will likewise be corrected.

We have completed our discussion of the classical Preisach model by showing its definition, its derivation, its identification techniques, and its properties. It works surprisingly well, considering its limitations. The hysteresis loops that it predicts, for nonsingular Preisach functions, have unit squareness, and in the next chapter we add a reversible component to remove this limitation.

2.8 CONCLUSIONS

In this chapter we presented the classical Preisach model. It describes a hysteresis loop with four variables: M_S, $\bar{h}_k$, σ_i, σ_k. The parameter M_S transforms the normalized Preisach loop into one whose height matches the magnetization of the medium. The parameter $\bar{h}_k$ determines the value of the coercivity and for Gaussian and other symmetrical functions is equal to the coercivity. The parameter $\sigma = (\sigma_i^2 + \sigma_k^2)^{0.5}$ determines the slope of the hysteresis loop at the coercivity, and the ratio of σ_i/σ_k determines the height of minor loops vis-a-vis the major loop.

We have shown explicitly that the Preisach model, for the case of a Gaussian Preisach function, computes a different slope and hence a different curve when the input is increasing from the slope and curve when the input is decreasing. Furthermore, for increasing inputs, the effect of history is contained in the last undeleted minimum, and for decreasing inputs, the history is contained in the last undeleted maximum. As each minimum is deleted, the slope is discontinuous.

The classical Preisach model creates minor loops that have the congruency and the deletion properties. The magnetization changes computed by it are irreversible. To characterize real magnetic materials, in Chapter 3 we will add reversible magnetization, in Chapter 4 we will relax the congruency property, in Chapter 5 we will relax the deletion property, and in Chapter 6 we will discuss vector properties.

REFERENCES

[1] F. Vajda and E. Della Torre, "Ferenc Preisach, In Memoriam," *IEEE Trans. Magn.* **MAG-31**, March 1995, pp. i–ii.

[2] F. Preisach, "Über die magnetische Nachwirkung," *Z. Phys.*, **94**, 1935, pp. 277–302.

[3] I. D. Mayergoyz, *Mathematical Models of Hysteresis,* Springer-Verlag: New York, 1991.

[4] F. Vajda and E. Della Torre, "Efficient numerical implementation of complete-moving-hysteresis models," *IEEE Trans. Magn.*, **MAG-29**, March 1993, pp. 1532–1537.

[5] E. P. Wohlfarth, "Relations between different modes of acquisition of the remanent magnetization of ferromagnetic particles," *J. Appl. Phys.*, **29**, March 1958, pp. 595–596.

[6] J. G. Woodward and E. Della Torre, "Particle interaction in magnetic recording tapes," *J. Appl. Phys.*, **31**, January 1960, pp. 56–62.

[7] E. Della Torre and F. Vajda, "The identification of the switching field distribution components," *IEEE Trans. Magn.*, **MAG-31**, September 1995, pp. 2536–2542.

CHAPTER 3

IRREVERSIBLE AND LOCALLY REVERSIBLE MAGNETIZATION

3.1 INTRODUCTION

This chapter deals with the first of the corrections to the classical Preisach model, the introduction of reversible magnetization, so that the model can describe magnetization phenomena accurately. Although the classical Preisach model can describe reversible magnetization, it is limited to a state-independent description. In magnetization-dependent models, the susceptibility can be a function of the applied field, but is independent of the magnetization. In state-dependent models, the susceptibility can be a function of both the field and the magnetization. Thus, the various models can give increasingly accurate descriptions of the reversible magnetization in real media.

3.2 STATE-INDEPENDENT REVERSIBLE MAGNETIZATION

The magnetization changes of a classical Preisach hysteron located in this *physically realizable region* of the Preisach plane, that is, the region in which $u > v$, are totally irreversible. By this we mean that the energy transfer associated with this change is not recoverable. We will refer to the component of the magnetization associated with this process as the *irreversible magnetization*, M_i. Examination of the behavior of a single hysteron, such as the Stoner–Wohlfarth particle in Fig. 1.11, shows that the magnetization can change reversibly as long as the appropriate critical field is not exceeded. This type of behavior can be characterized by adding a reversible component to the magnetization.

Energy transferred by an applied field to the reversible component is stored and can be totally recovered when the applied field is returned to zero. Thus, the *reversible magnetization*, M_r, is given by a single-valued function of the applied field,

$$M_r(H) = F(H). \tag{3.1}$$

This single-valued function has the property that $F(0)$ is zero and $F(\infty)$ is finite. Since there is no preferred direction of magnetization,

$$F(\infty) = -F(-\infty). \tag{3.2}$$

This type of behavior could be characterized by hysterons on the $u = v$ diagonal. The Preisach function necessary to achieve this magnetization is

$$P(U,V) = M_r(U)\ \delta(U-V) = F(U)\ \delta(U-V), \tag{3.3}$$

where δ is the Dirac delta function, which is zero unless its argument is zero but whose integral is unity; however, we will simply add the function, $F(U)$, to the Preisach integral to obtain the total magnetization. Then the total magnetization, or what we will simply call the magnetization, is given by

$$M = M_i + M_r. \tag{3.4}$$

The remanence, M_{rem}, is the magnetization when the applied field is zero. Since the reversible magnetization M_r is zero, when H is equal to zero, then M_{rem} is equal to the M_i at zero field for single-quadrant media.

We define the squareness, S, of a material to be the ratio of the maximum remanence to the maximum magnetization. Then in terms of normalized magnetizations, we have

$$M(H) = M_S[Sm_i(H) + (1-S)m_r(H)] = M_S[Sm_i(H) + (1-S)f(H)]. \tag{3.5}$$

This functionality is illustrated by the block diagram in Fig. 3.1.

The normalized reversible magnetization is defined to be 1 as the applied field approaches infinity. Furthermore, since the material does not have any preferred direction of magnetization, the reversible magnetization must be an odd function of the applied field. Thus, $f(0)$ and all even derivatives of m_r, at zero applied field, have to be zero, and since the reversible magnetization must saturate, the second derivative must decrease with increasing H. With this definition, the magnitude of m is less than one, and the normalized function, $f(H)$, as defined by

$$f(H) = \frac{F(H)}{(1-S)M_S}, \tag{3.6}$$

has the following properties:

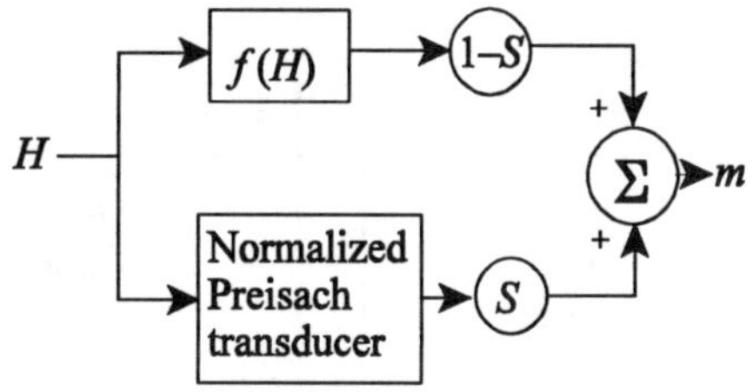

Figure 3.1 Block diagram of a Preisach transducer with state-independent reversible magnetization.

$$f(x) = -f(-x), \quad f(0) = 0, \quad \text{and} \quad f(\infty) = 1. \tag{3.7}$$

It is seen from the Stoner–Wohlfarth model, (Fig. 1.11), that when the hysteron is in its positive state, its reversible susceptibility, dm_r/dH, is a monotonically decreasing function of H, for all H greater than $-H_S$. Similarly in its negative state, dm_r/dH is a monotonically increasing function of H, for all H less than H_S. Thus, the reversible magnetization has to be either magnetization dependent or state dependent, as shown in the following sections.

This type of behavior is *state independent*, since the reversible component depends only upon the applied field. The next section discusses *magnetization-dependent* and *state-dependent models*.

3.3 MAGNETIZATION-DEPENDENT REVERSIBLE MODEL

The DOK model [1], a magnetization-dependent model, assumes that the reversible magnetization depends upon the magnetization state of the hysterons. Let the reversible magnetization when the hysteron is in the positive state be $f(H)$. Then, if a_+ and a_- are the fractions of hysterons in the positive and negative states, respectively, the reversible magnetization is given by

$$m_r = a_+ f(H) - a_- f(-H). \tag{3.8}$$

With this definition of a reversible component, we can remove the restriction for large negative fields; hence, the function f is restricted only by

$$f(0) = 0 \quad \text{and} \quad f(\infty) = 1. \tag{3.9}$$

The decomposition of a hysteron's loop into an irreversible component and a magnetization-dependent reversible component is shown in Fig. 3.2. It is seen that the a's are given by

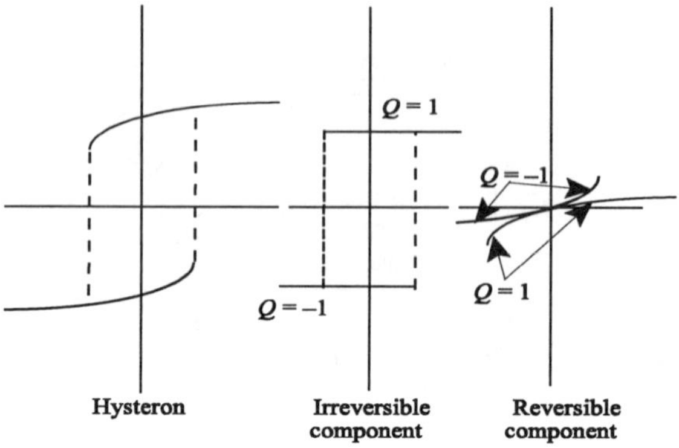

Figure 3.2 Decomposition of hysteron into irreversible and locally reversible components.

$$a_+ = \frac{1+m_i}{2} \tag{3.10}$$

and

$$a_- = \frac{1-m_i}{2}. \tag{3.11}$$

Thus,

$$a_+ + a_- = 1. \tag{3.12}$$

A block diagram of the resulting model is shown in Fig. 3.3. With this model, the reversible magnetization is now *magnetization dependent.* This type of reversible magnetization changes abruptly when the state of the hysteron changes, so we call it *locally reversible magnetization.* In this case, since for large negative fields, m_i approaches −1, consequently a_+ approaches zero. Thus, there is no restriction on how $f(H)$ behaves for large negative values of H. It follows that unlike the case of the magnetization-dependent reversible magnetization, $f'(H)$ could be a monotonically decreasing function of H, and $f''(H)$ could be negative for all H, since neither contributes to the magnetization for large negative values of H. The only restrictions on $f(H)$ are that it approaches one as H goes to infinity, and that it is zero when H is zero. Since

$$\chi_r(H,M) = (1-S)M_S\frac{dm_r}{dH} = (1-S)M_S[a_+f'(H) + a_-f'(-H)], \tag{3.13}$$

we have

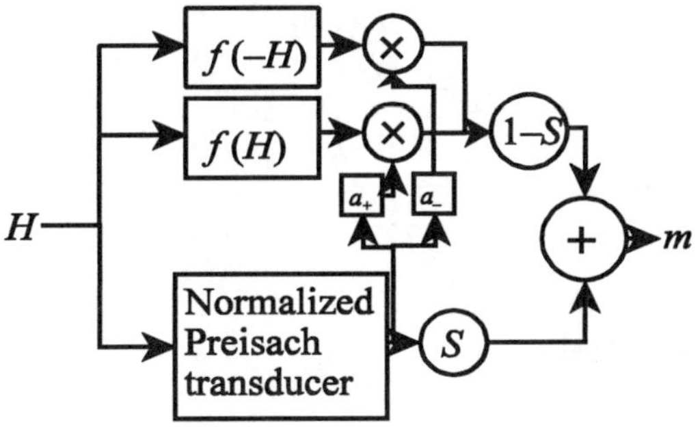

Figure 3.3 Preisach model with state-dependent reversible magnetization.

$$\chi_r(0) = (1-S)\,M_S\,f'(0). \tag{3.14}$$

This is independent of m_i, which is where the magnetization curve crosses the axis. Since in many materials this property is not present, we will examine this in more detail in the next section.

For a collection of Stoner–Wohlfarth particles, $f(H)$ should be the normalized reversible component of the magnetization curve. It is useful to approximate this function by

$$f(H) = 1 - e^{-\xi H}. \tag{3.15}$$

This approximation is illustrated in Fig. 3.4 as compared to the Stoner–Wohlfarth model, and in Fig. 3.5, as compared to γ-Fe_2O_3 data. Although the Stoner–Wohlfarth fits the measurements better than the exponential, the error is not large.

Since the susceptibility is given by

$$\chi = \frac{dM}{dH} = (1-S)\,M_S\,f'(h) = (1-S)\,M_S\,\xi e^{-\xi H}, \tag{3.16}$$

then

$$\xi = \frac{\chi_0}{(1-S)M_S}, \tag{3.17}$$

where χ_0 is the zero-field susceptibility.

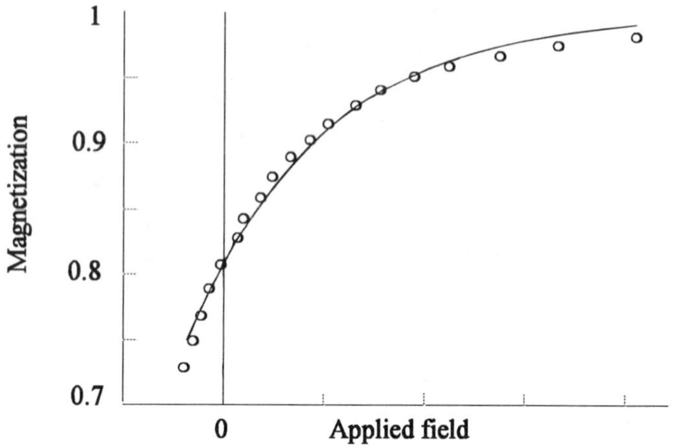

Figure 3.4 Comparison of the exponential approximation (circles) with the Stoner–Wohlfarth model (solid line).

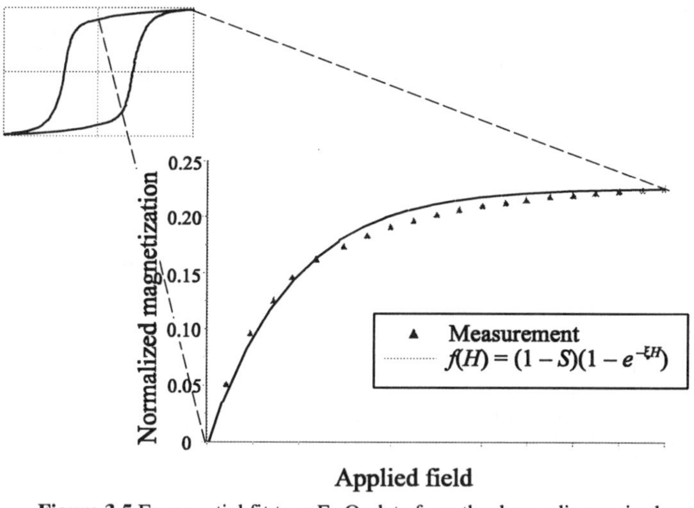

Figure 3.5 Exponential fit to γ-Fe_2O_3 data from the descending major loop.

3.4 STATE-DEPENDENT REVERSIBLE MODEL

Hysteron loops of isolated hysterons have to be symmetrical with respect to the origin; consequently, we can attribute the asymmetry required in the Preisach model to interaction between hysterons. Thus, we say that the field that a hysteron sees is the sum of the applied field and the interaction field. The effect of this field

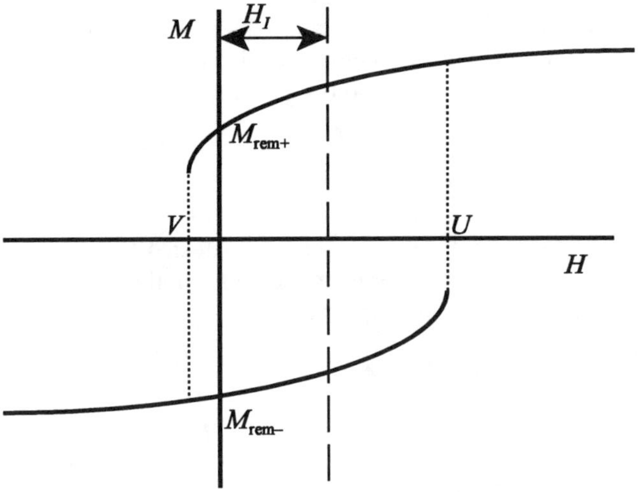

Figure 3.6 Hysteron in the presence of an interaction field.

is to displace the hysteresis loop horizontally, as shown in Fig. 3.6 for a material that has locally reversible magnetization. It is seen that when this happens, not only does U not equal the negative of V, but the positive remanence, M_{rem+}, does not equal the negative of the negative remanence, M_{rem-}. This difference is taken into account by the CMH model [2], which is a state-dependent magnetization model.

It is seen that the new values of the remanence are given by

$$m_{r+} = 1 + f(H_i), \tag{3.18}$$

and

$$m_{r-} = -1 + f(-H_i), \tag{3.19}$$

where H_i is the value of the interaction field and f is the same type of function discussed in section 3.3. Then, for positive hysterons with an average squareness S_A, the magnetization is given by

$$m(H) = S_A m_i + (1 - S_A)\, f(H + H_i). \tag{3.20}$$

Then, summing over all hysterons, we obtain

$$m = \int_{H_k>0} \int Q(H_i, H_k) p(H_k, H_i) \left\{ S_A + (1 - S_A) f[Q(H_k, H_i)(H + H_i)] \right\} dH_k dH_i, \tag{3.21}$$

where, as before, Q is +1 in the region that is magnetized positively and –1 otherwise. The irreversible component m_i changes only when there is a change of state. The reversible component changes with the applied field in a manner that depends on the state and is zero when the applied field is zero. Thus, we can solve for m_i by setting H equal to zero in this equation. Then we get

$$m_i = \int_{H_k>0}\int Q(H_k,H_i)p(H_k,H_i)\{S_A + (1-S_A)f[H_i\ Q(H_k,H_i)]\}\ dH_k dH_i. \tag{3.22}$$

We see that in this model, the remanence is affected by this correction in the second term inside the braces. For example, at saturation Q is unity everywhere, and this reduces to the observed squareness S, which is now

$$\begin{aligned} S &= \int_{H_k>0}\int p(H_k,H_i)\{S_A + (1-S_A)f(-H_i)\}dH_k dH_i \\ &= S_A + (1-S_A)\int_{H_k>0}\int p(H_k,H_i)f(-H_i)\,dH_k dH_i. \end{aligned} \tag{3.23}$$

We see that if the average value of $f(-H_i)$ were zero, the observed squareness would be the same as the average squareness; however, for real materials there is a correction given by the second term.

When the irreversible magnetization is subtracted from the total magnetization, (3.21), the remainder is the locally reversible component

$$\begin{aligned} m_r = (1-S)\Big\{ &\int_{Q=1}\int p(H_k,H_i)[f(H+H_i) - f(H_i)]dH_k dH_i \\ &+ \int_{Q=-1}\int p(H_k,H_i)[f(H_i) - f(H+H_i)]dH_k dH_i\Big\}. \end{aligned} \tag{3.24}$$

We note that if (3.15) is applied, then

$$f(H+H_i) - f(H_i) = e^{-\xi(H+H_i)} - e^{-\xi H_i} = e^{-\xi H_i}\left[e^{-\xi H} - 1\right], \tag{3.25}$$

so that we can write

$$m_r = a_+ f(H) - a_- f(-H), \tag{3.26}$$

where now

$$a_+ = \int_{Q=1}\int \exp(-\xi H_i)\ p(H_k,H_i)\ dH_k dH_i, \tag{3.27}$$

and

$$a_{-} = \int_{Q=-1}\int \exp(\xi H_i)\ p(H_k, H_i)\ dH_k dH_i. \tag{3.28}$$

This model is now *state dependent*, since even with the same magnetization, different values can be obtained for a_+ and a_-. A major difference between this model and the preceding model is that a_+ and a_- no longer have to add up to one. Thus, the susceptibility is now a function of the magnetic state; hence the zero-field susceptibility depends on the magnetization and how that magnetization was achieved.

To illustrate the effect of this model, let us consider the variation of the susceptibility along the M axis for a *dc* magnetizing process using a Gaussian Preisach function. In that case, a_+ is given by

$$a_{+} = \frac{1}{2\pi\sigma_c\sigma_i}\int_{Q=1}\int \exp(-\xi H_i)\exp\left\{-\frac{1}{2}\left[\left(\frac{H_k - \bar{h}_k}{\sigma_k}\right)^2 + \left(\frac{H_i}{\sigma_i}\right)^2\right]\right\} dH_i\ dH_k$$

$$= \frac{\exp\left(\frac{\xi^2\sigma_i^2}{2}\right)}{2\pi\sigma_k\sigma_i}\int_{Q=1}\int \exp\left\{-\frac{1}{2}\left[\left(\frac{H_k - \bar{h}_k}{\sigma_k}\right)^2 + \left(\frac{H_i}{\sigma_i} + \xi\sigma_i\right)^2\right]\right\} dH_i\ dH_k. \tag{3.29}$$

For a *dc* magnetizing process starting from negative saturation going to a positive field H_1 and then returning to zero, we have

$$a_{+} = \exp\left(\frac{\xi^2\sigma_i^2}{2}\right)\operatorname{erf}\left(\frac{H_1 - \bar{h}_k + \xi\sigma_i^2}{\sigma}\right) \tag{3.30}$$

and

$$a_{-} = \exp\left(\frac{\xi^2\sigma_i^2}{2}\right)\left[1 - \operatorname{erf}\left(\frac{H_1 - \bar{h}_k - \xi\sigma_i^2}{\sigma}\right)\right]. \tag{3.31}$$

It is seen that if $\xi\sigma_i^2$ is zero, then the sum of the a's is again unity. Since $\xi\sigma_i^2$ is always positive, the two functions overlap, as indicated in Fig. 3.7. The resulting zero-field susceptibility, as shown in the figure, is largest at the coercive field and approaches $\exp(\xi^2\sigma_i^2/2)$ as the magnitude of H increases. Since the remanence is a single-valued function of the applied field, the susceptibility as a function of the remanence has a similar shape that increases to a maximum at zero remanence and then decreases. This is generally similar to the observed susceptibility in recording media [3]. The main difference between this calculation and the observation is that the observed peak in susceptibility does not occur for zero magnetization. This discrepancy can be explained by the moving model, which is discussed in the next chapter.

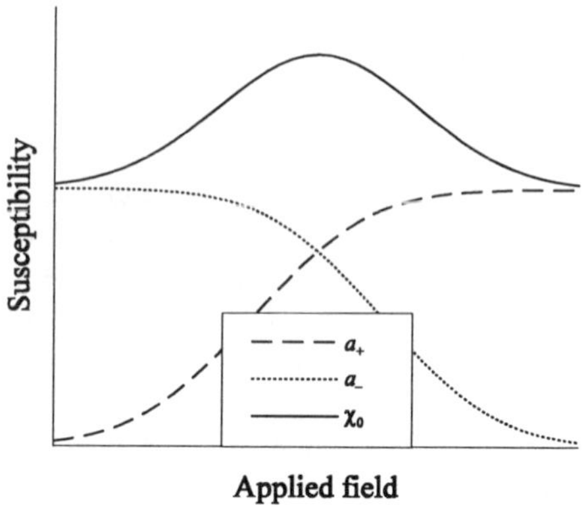

Figure 3.7 Variation of susceptibility with magnetization for a *DC*-magnetizing process.

3.5 ENERGY CONSIDERATIONS

It is well known that the energy density lost in a closed magnetization cycle is equal to the area enclosed by the *M*-*H* loop. Friedman and Mayergoyz [4] have suggested that the energy lost in an open process can be computed from the classical Preisach model. They later extended this analysis to input-dependent Preisach models [5]. We now address the question of energy storage and dissipation of the models discussed in the preceding sections. The irreversible component dissipates energy every time it changes, and it is incapable of storing energy. The locally reversible component, on the other hand, stores an amount of energy that depends on the state of the system. Since the energy stored varies when the system changes state, even if the applied field is unchanged, this change in energy must be added or subtracted from energy dissipated by the irreversible component. This fact complicates the computation of the energy loss for an open cycle.

The energy relations for a single hysteron can be obtained by examining the hysteresis loop for an isolated hysteron, as shown in Fig. 3.8. The magnetization M, the solid curve, can be decomposed into the sum of a reversible component, M_r, illustrated by the two dashed curves, and an irreversible component M_i, illustrated by the rectangular hysteresis loop. That is,

$$M(H) = M_i(Q) + M_r(H,Q), \tag{3.32}$$

where Q is the state of the hysteron. The functional variation of the reversible component for the magnetization $f(H)$ is a concave, monotonic, single-valued function of the applied field, which saturates as H approaches infinity.

We will now compute w, the energy dissipated in going from zero applied field to H_k, and back to zero. This is equal to μ_0 times the area between the hysteresis loop and the M axis and can be written

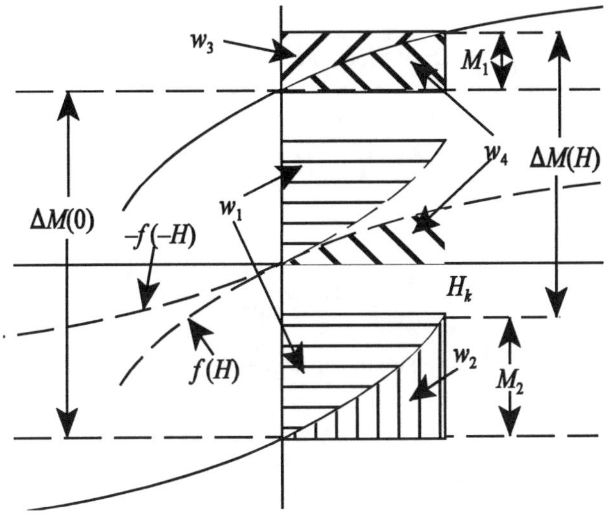

Figure 3.8 Hysteresis loop of an isolated hysteron.

$$w = \mu_0[\Delta M(0)H_k + w_4 - w_3]. \tag{3.33}$$

It is seen that the hatched area of the rectangle at the lower right-hand corner of the hysteresis loop, is given by

$$M_2 H_k = w_1 + w_2, \tag{3.34}$$

and the area of the hatched rectangle at the upper right hand corner of the hysteresis loop is given by

$$M_1 H_k = w_3 + w_4. \tag{3. 35}$$

The discontinuity in the hysteresis loop when the hysteron changes state, $\Delta M(H_k)$, can be obtained from the height of the irreversible loop at its center, $\Delta M(0)$, as

$$\Delta M(H_k) = \Delta M(0) + M_1 - M_2. \tag{3.36}$$

Thus, (3.33) becomes

$$w = \mu_0[\Delta M(H_k)\ H_k + w_1 - w_3]. \tag{3.37}$$

The first term is the energy loss corresponding to the discontinuity in the magnetization, while the remaining terms correspond to the change in the energy stored in the locally reversible magnetization.

The energy stored in the locally reversible magnetization, w_r, is

$$w_r = \mu_0 \int_0^{M_r} H dM_r|_{Q=\text{const.}} - \mu_0 \left[H\ M_r - \int_0^H M_r dH|_{Q=\text{const.}}\right]. \tag{3.38}$$

Since

$$M_1 = f\,(H_k) \quad \text{and} \quad M_2 = -f\,(-H_k), \tag{3.39}$$

then w_1 and w_3 can be written as

$$w_1 = w_R\big|_{Q=-1} = -H_k\, f(-H_k) + \int_0^{H_k} f(-H)\; dH, \tag{3.40}$$

and

$$w_3 = w_R\big|_{Q=1} = H_k\, f(H_k) - \int_0^{H_k} f(H)\, dH. \tag{3.41}$$

Therefore, the energy dissipated in traversing the left half of the hysteresis loop is given by

$$w = \mu_0 \left\{ H_k [M_j - f(H_k) - f(-H_k)] + \int_0^{H_k} [f(H) + f(-H)] dH \right\}. \tag{3.42}$$

The first term is the magnetization change in the irreversible component and is equal to the product of the coercivity and the size of the Barkhausen jump. The second two terms are due to the magnetization change in the reversible component, and the integral is the change in stored energy in the reversible component.

3.5.1 Hysteron Assemblies

We will now generalize this result for an assembly of hysterons by essentially summing this over all the hysterons. The result will be a generalization of (3.42). The energy supplied to a magnetic medium is given by

$$W = \mu_0 \int_0^M H\, dM = \mu_0 \int_0^H H \frac{dM}{dH} dH. \tag{3.43}$$

If the magnetization M, is the sum of an irreversible component M_i, and a locally reversible component M_r, then the rate of change in the magnetization with respect to the applied field is

$$\frac{dM}{dH} = \frac{dM_i}{dQ} + \frac{\partial M_r}{\partial Q}\frac{dQ}{dH} + \frac{\partial M_r}{\partial H}. \tag{3.44}$$

The first two terms correspond to irreversible changes in the magnetization with respect to the applied field and, therefore, are a source of dissipation. There is an additional dissipation term due to the changing ability of the medium to store energy in the different irreversible states for the same applied field. The energy stored in the reversible component for a given state, W_S, is given by

$$W_S = \mu_0 \int_0^H U \frac{\partial M_r(U,Q)}{\partial U} dU, \tag{3.45}$$

where U is a dummy variable of integration. Thus, the rate of increase in the energy stored in the reversible component is given by

$$\frac{dW_S}{dH} = \mu_0 H \frac{\partial M_r}{\partial H}, \tag{3.46}$$

and the dissipated energy, W_D, in the medium is given by

$$\frac{dW_D}{dH} = \frac{\partial W_S}{\partial Q}\frac{dQ}{dH} + \int_0^H U\left[\frac{dM_r}{dU} + \frac{\partial M_r}{\partial Q}\frac{dQ}{dU}\right] dU. \tag{3.47}$$

If, furthermore, the reversible component can be factored, as in (3.15), then a specific formula for the energy dissipated can be derived. The reversible component of the magnetization at the jth element is then given by

$$M_{r,j} = \pm g(\pm H_{i,j})\, f(\pm H), \tag{3.48}$$

where the upper sign is to be used if the hysteron is in the upper magnetization state, H is the applied field, and $H_{i,j}$ is the interaction field at the jth hysteron. The rate of increase of this reversible magnetization with the applied field is given by

$$\frac{dM_r}{dH} = \frac{\partial M_r}{\partial a_+}\frac{da_+}{dH} + \frac{\partial M_r}{\partial a_-}\frac{da_-}{dH} + \frac{\partial M_r}{\partial H} = f(H)\frac{da_+}{dH} - f(-H)\frac{da_-}{dH} + \frac{\partial M_r}{\partial H}, \tag{3.49}$$

where the first two terms are the change in M_r due to a change in state, and the last term is the change in M_r due to the change in the applied field. The last term is given by

$$\frac{\partial M_r}{\partial H} = a_+ \frac{df(H)}{dH} + a_- \frac{df(-H)}{dH}, \tag{3.50}$$

and the derivatives of the a's are given by

$$\frac{da_\pm}{dH} = \pm(1-S)\int_{H_1}^{H} g\left(\pm\frac{H+H_-}{2}\right) P(H,H_-)\, dH_-, \tag{3.51}$$

where H_1 is defined in (2.30). It is noted that if $g(u,v)$ is zero outside the fourth quadrant, the derivatives are zero when the magnitude of H is decreasing, and W_S is the recoverable energy.

Thus, the rate of energy dissipation is given by

$$\frac{dW}{dH} = \mu_0 H \left[\int_{H_1}^{H} P(H,H_+)dH_+ + \frac{\partial a_+}{dH} f(H) - \frac{\partial a_-}{dH} f(-H) \right] + \int_0^H U \left[\frac{\partial a_+}{dH} f(U) + \frac{\partial a_-}{dH} f(-U) \right] dU. \tag{3.52}$$

The first term corresponds to the rate of energy dissipation due to the discontinuous change in the irreversible magnetization; the second term corresponds to the discontinuous change in the reversible magnetization when the system changes state; and the last term is the change in the energy stored in the reversible magnetization due to changes in the state of the system.

The energy dissipated by a magnetic material has been computed for both a single hysteron and an assembly of hysterons represented by a state-dependent Preisach model. For an isolated hysteron, the two components of the energy loss are due to the sudden change in irreversible magnetization and due to the change in the ability of the magnetization to store energy because of the change in state. The former loss is equal to the product of the permeability, the applied field, and the discontinuous magnetization change. It is noted that the size of the discontinuous change in magnetization is generally not equal to the difference of the two remanent magnetizations. These energy loss components carry over into the computation for an assembly of hysterons. The reversible component is state independent, if and only if both the sum of the a_+ and a_- is constant, and $f(H)$ is an odd function: that is, $f(H) = -f(-H)$.

3.6 IDENTIFICATION OF MODEL PARAMETERS

In this section, we will limit our attention to the identification of the additional parameters necessary to identify the reversible component of magnetization for single-quadrant media. In this case, the identification of the irreversible component of magnetization for any point in the M-H plane can be performed by reducing the applied field to zero. Then, the reversible component is reduced to zero, and since the medium is single quadrant, there is no change in the irreversible component. In the next section, we will discuss the problem encountered in three-quadrant media.

For all these media, to completely identify the reversible component one must measure the squareness S and the function $f(H)$. The squareness is simply the measured ratio of the maximum remanence to the saturation magnetization. The method of measuring $f(H)$ depends on the model.

For state-independent reversible magnetization, one must find the function that describes the reversible component as a function of the applied field. This can be measured directly by simply applying the field and reducing it to zero. The accuracy of the measurement is determined by the accuracy with which one can set a field and measure the magnetization. If observed variation is approximated by a function such as a hyperbolic tangent or a Gudermannian, the few parameters associated with these functions can be obtained by a technique such as curve fitting. The validity of the model could be determined by simply seeing if the measured value of $f(H)$ is indeed independent of the magnetizing process.

For the magnetization-dependent models, the function $f(H)$ can be obtained for positive values of H from the descending major remanence loop directly. For negative values of H, the function can be obtained, once $f(H)$ has been obtained for positive H, by measuring the reversible component and substituting into

$$f(-H) = \frac{a_+ f(H) - m_r}{a_-}, \tag{3.53}$$

since a_+ and a_- can be determined directly from the measurement of m_i.

The function for the state-dependent reversible component is the most complicated to obtain, since it depends on the shape of the Preisach function as well as the magnetizing process. Furthermore, the observed squareness differs from the average squareness of the hysterons as computed in (3.23). Although it is possible to perform the necessary integrations numerically and obtain the functional variation directly, it is preferable to approximate the Preisach function appropriately by a function such as the Gaussian, and the reversible magnetization by an exponential, as discussed earlier. In that case, the process is described by two parameters, S and χ_0.

3.7 APPARENT REVERSIBLE MAGNETIZATION

The preceding discussion of reversible magnetization models was limited to single-quadrant media, that is, materials with moments sufficiently smaller than the coercivity to confine the only significant portion of the Preisach function to the fourth quadrant of the Preisach plane. On the other hand, media with larger moments can have a standard deviation of the interaction field, σ_i, large enough to allow the Preisach function to spill outside the fourth quadrant. High moment materials, such as Co-Cr-Ta, typical of media used in hard drives, under certain circumstances can have three-quadrant Preisach functions.

For three-quadrant media, one must distinguish between the remanence and the irreversible component of the magnetization. The remanence is the measurable magnetization when the applied field is removed, whereas the irreversible component of the magnetization usually is not measurable, but merely a convenient component of the decomposition of the total magnetization. These two quantities are identical for *single-quadrant* media, and the former has been used in experiments as an estimate for the latter, since the locally reversible component of the magnetization is zero in the absence of an applied field. The difference between the remanence and the state-dependent irreversible component of the magnetization is called the *apparent reversible effect* [6].

Preisach hysterons that lie in the first or third quadrant are hysteretic, although they have a unique state when the applied field is zero. For example, first-quadrant hysterons have both up- and down-switching fields that are positive, so in the absence of an applied field they are always magnetized negatively. The magnetization of such hysterons subtract from the maximum possible positive remanence. Since they traverse a hysteresis loop, whenever the applied field is cycled between zero and a value larger than its up-switching field, they will dissipate energy. This is not to be confused with the reversible component of magnetization, which does not dissipate energy as long as the magnetization state does not change. Thus, in the first quadrant of the hysteresis loop, the irreversible component of magnetization for decreasing applied fields is no longer horizontal for these materials and is not directly measurable.

We will now compute the correction to the descending major remanence curve. For convenience, we will extend the definition of the descending remanence curve to positive fields by setting it equal to the remanence at zero applied field, when the applied field is positive. Then, the apparent reversible magnetization, m_{AR}, can be defined for all H as

$$m_{AR} = m_i(H) - m_r(H), \tag{3.54}$$

where m_r is the remanence. When the remanence is computed after a positive field has been applied, the irreversible component of the magnetization must be reduced by the integral of the Preisach function over the region with the vertical hatching in Fig. 3.9. On the other hand, for negative fields, it must be increased by the integral of the Preisach function over the region with the horizontal hatching in that figure.

For the descending major loop, we can get the variation in irreversible magnetization by setting H_1 equal to negative infinity in (2.31). Thus,

$$\chi_{AR} = \frac{dm_i(H)}{dH} = \frac{1}{\sigma}\sqrt{\frac{2}{\pi}}\exp\left[-\frac{(H+\bar{h}_k)^2}{2\sigma^2}\right] \quad \text{for } H>0. \tag{3.55}$$

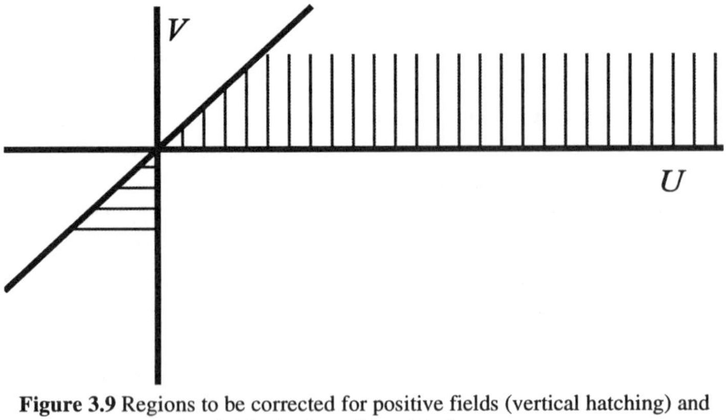

Figure 3.9 Regions to be corrected for positive fields (vertical hatching) and negative fields (horizontal hatching).

We see that at H equal to zero, the slope of m_i is given by

$$\frac{dm_i(0)}{dH} = \frac{1}{\sigma}\sqrt{\frac{2}{\pi}}\exp\left[-\frac{H_k^2}{2\sigma^2}\right], \tag{3.56}$$

which is not zero as it is in the case of single quadrant media. It was shown in general that $m_i(H)$ for the descending major loop is given by

$$m_i(H) = \mathrm{erf}\left(\frac{H+\bar{h}_k}{\sigma\sqrt{2}}\right). \tag{3.57}$$

Thus, when there is no reversible magnetization, the squareness due to apparent reversible magnetization of this medium, S_A, is given by

$$S_A = \mathrm{erf}\left(\frac{\bar{h}_k}{\sigma\sqrt{2}}\right). \tag{3.58}$$

If the material has in addition a reversible component of magnetization, it must be added to m_i, as before. If the squareness due to m_r is called S_r, then the squareness of this material is given by

$$S = S_A S_r. \tag{3.59}$$

For positive applied fields, the descending major remanence loop is a constant given by S_A. Therefore, the magnetization due to apparent reversible magnetization, m_{AR}, the vertically hatched region of Fig. 3.9, is the difference of between S_A and $m_i(H)$. That is,

$$m_{AR}(H) = \mathrm{erf}\left(\frac{\bar{h}_k}{\sigma\sqrt{2}}\right) \quad \mathrm{erf}\left(\frac{H+\bar{h}_k}{\sigma\sqrt{2}}\right). \tag{3.60}$$

For negative applied fields, the remanence loop is obtained by adding m_{AR}, the contribution of the horizontally hatched region of Fig. 3.9, to $m_i(H)$. We see that

$$m_{AR} = \int_H^0 \chi_{AR}(H)\,dH, \tag{3.61}$$

where H is a negative number. We can obtain χ_{AR} by substituting zero for H_1 in the negative of (2.30). Thus,

$$\chi_{AR} = \frac{dm_{AR}}{dH}$$
$$= \frac{1}{\sigma\sqrt{2\pi}} \exp\left[-\frac{(H+\bar{h}_k)^2}{2\sigma^2}\right]\left[\operatorname{erf}\left(\frac{\lambda H + \kappa\bar{h}_k}{\tau\sqrt{(2)}}\right) - \operatorname{erf}\left(\frac{(1+\lambda)H + \kappa\bar{h}_k}{\tau\sqrt{(2)}}\right)\right], \tag{3.62}$$

for $H < 0$. It is noted that χ_{AR} is zero when H is zero.

A plot of χ_{AR} as a function of the applied field is shown in Fig. 3.10. The applied field is normalized to the coercivity, σ_i was taken to be four times the coercivity, and σ_k was taken to 0.4 times the coercivity. It is seen that the susceptibility is always positive.

The effect of apparent susceptibility can be seen by examining Fig. 3.11. With this set of parameter values, the apparent squareness S_A is 0.2. It is seen that the remanence is constant for positive fields and decreases with decreasing negative fields. Furthermore, the slope of the remanence is positive for small negative fields, an indication of substantial apparent reversible magnetization. Most important, the irreversible magnetization is distinctly different from the remanence. Also, the remanence coercivity, H_{RC}, is not a good measure of $\bar{h}_k$. For these values of parameters, the remanent coercivity is only 8% greater than the mean critical field. This slightly complicates the identification problem, as we will see in the next chapter.

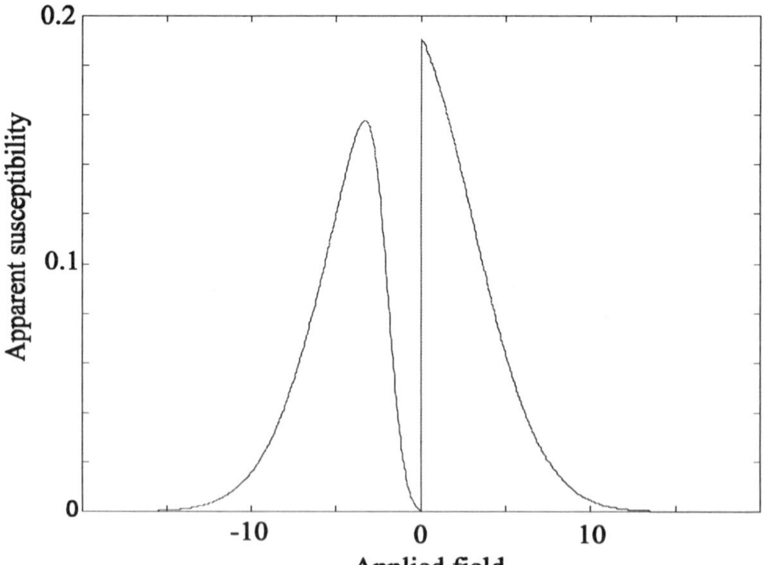

Figure 3.10 Variation in apparent susceptibility with applied field.

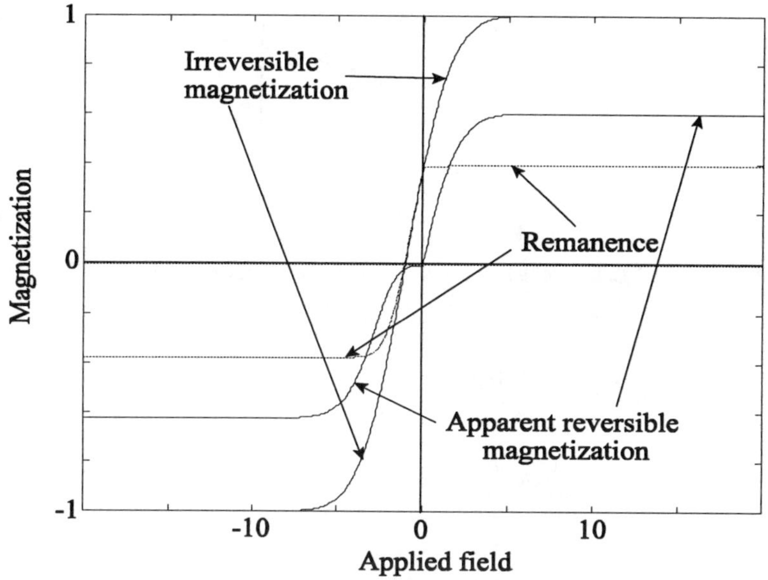

Figure 3.11 Effect of apparent reversible magnetization on remanence.

3.8 CROSSOVER CONDITION

There is a limit on the choice of parameters that will produce a physically realizable model. In the case of state-independent reversal models, the only limits on the functions are that the Preisach function be zero if $U < V$ and that the reversible component be a monotonic single-valued function so that the material does not violate conservation of energy. For magnetization-dependent and state-dependent models there is another condition that all physically realizable hysterons must satisfy, the so-called the *crossover condition* [7], which limits the range of possible parameters permitted where the Preisach function is nonzero. Let us examine Fig. 3.12, where the value of the critical field was chosen to be too large for the values of the squareness and the zero-field susceptibility. In that case, the part of the hysteresis loop that is above the crossover point is traversed in the clockwise sense, violating conservation of energy.

We can write the magnetization as

$$m(H) = \begin{cases} S+(1-S)f(H), & \text{if } Q=1 \\ -S-(1-S)f(-H), & \text{if } Q=-1. \end{cases} \tag{3.63}$$

In order to avoid crossovers, we require that for all hysterons we have

$$m(H)|_{Q=1} \geq m(H)|_{Q=-1}, \quad \forall H \text{ such that } V<H<U. \tag{3.64}$$

Thus, for a particular hysteron we must have

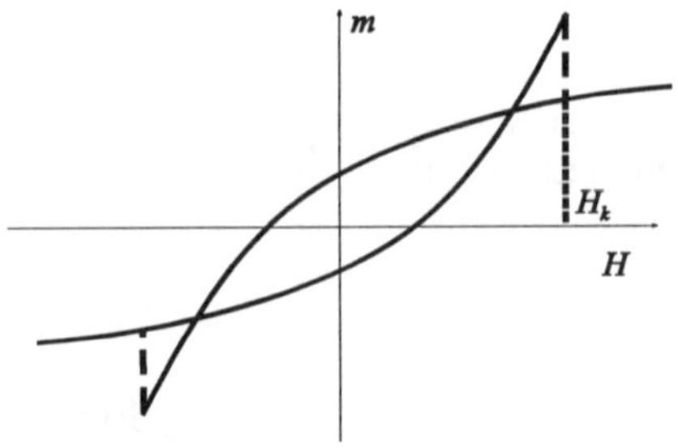

Figure 3.12 Loop of a hysteron that violates the crossover condition.

$$2S + (1-S)[f(H_k) + f(-H_k)] \geq 0, \tag{3.65}$$

or

$$\frac{f(H_k) + f(-H_k)}{2} \geq -\frac{S}{1-S}. \tag{3.66}$$

This puts a lower bound on the permissible squareness or an upper bound on the permissible zero-field susceptibility, depending on the position on the Preisach plane. Note that there is no limit in the case of $S = 1$. We can interpret this as the maximum permissible extent of the Preisach function on the Preisach plane, if S is uniform over the Preisach plane. Alternately, if S is permitted to be a function of H_k, then (3.65) is the constraint on permissible functions. In that case, the functions used must also satisfy the condition that the average S must agree with the observed value. This is of course even more complicated if there is any apparent reversible component to the magnetization.

If we use the exponential variation for $f(H)$, as in the preceding sections, then

$$\frac{f(H_k) + f(-H_k)}{2} = 1 - \cosh(\xi H_k), \tag{3.67}$$

and (3.65) becomes

$$\cosh(\xi H_k) \leq \frac{1}{1-S}. \tag{3.68}$$

If we permit either S or ξ to be a function of H_k, it must satisfy, respectively

$$S \geq 1 - \operatorname{sech}(\xi H_k), \tag{3.69}$$

or

$$\xi \leq \frac{1}{H_k} \cosh\left(\frac{1}{1-S}\right). \tag{3.70}$$

Since the hyperbolic secant lies between zero and one, it is seen that this lower limit for S also lies between zero and one, and approaches one for large H_k.

We have seen that in order to obtain a realistic hysteresis loop, we have to add a reversible component to the Preisach model. This component may be a function of the applied field only, or may depend on the magnetization or the state as well. The consequences of the state dependence on the hysteresis loop was discussed. If the Preisach function is nonzero outside the fourth quadrant, the irreversible component of the magnetization will be different from the remanence. This complicates the identification problem and leads to apparent reversible behavior.

3.9 CONCLUSIONS

In this chapter we discussed how reversible magnetization can be added to a Preisach model. As a result two new parameters were added to the model: S and χ. The first is the fraction of the saturation magnetization due to irreversible magnetization, and the second is the saturation reversible susceptibility at zero field. Three types of reversible magnetization processes were discussed: magnetization-independent, magnetization-dependent and state-dependent reversible magnetization. Only the first of these could have been characterized by the classical Preisach model. For magnetization-dependent reversible processes, the reversible susceptibility is a function of the applied field only. For a magnetization dependent process the reversible susceptibility is a function of both the applied field and the magnetization; however, at zero field, it is a constant. Only for state-dependent reversible processes does the susceptibility vary at zero field.

It may be said that if one is only interested in computing the remanence, it is not necessary to compute the reversible component of the magnetization. This is true for the classical Preisach model, but not for the moving model and the product model discussed the next chapter. We will see that for these two models, even to compute the remanence, we must compute both the reversible and the irreversible components of the magnetization. Errors can be considerable in these cases, especially for soft magnetic materials, if one neglects the reversible component or does not include the correct variation of it.

REFERENCES

[1] E. Della Torre, J. Oti, and G. Kádár, "Preisach modeling and reversible magnetization," *IEEE Trans. Magn*, **MAG-26**, November 1990, pp. 3052–3058.

[2] F. Vajda and E. Della Torre, "Characteristics of magnetic media models," *IEEE Trans. Magn.*, **MAG-28,** September 1992, pp. 2611–2613.

[3] F. Vajda and E. Della Torre, "Reversible magnetization models for magnetic recording media," *Physica B*, **223**, June 1997, pp. 330–336.

[4] I. D. Mayergoyz and G. Friedman, "The Preisach model and hysteretic energy losses," *J. Appl. Phys.*, **61**, April 1987, 3910–3912.

[5] G. Friedman and I. D. Mayergoyz, “Input-dependent Preisach models and hysteretic energy losses,” *J. Appl. Phys.*, **69**, April 1991, pp. 4611–4613.

[6] O. Benda, “The question of the reversible processes in the Preisach model,” *Elect. Engg. J. Slovak Acad. Sci.,* **6**, 1991.

[7] F. Vajda and E. Della Torre, “Scalar characterization of magnetic recording media (invited),” *Nanophases and nanocrystalline structures*, R. D. Shull and J. M. Sanchez, eds. TMS: Warrendale, PA, 1993, pp. 121–133.

CHAPTER 4

THE MOVING MODEL AND THE PRODUCT MODEL

4.1 INTRODUCTION

So far, we have assumed that a Preisach function exists for a given magnetic material. In this chapter, we address the questions of why it should exist at all, whether it is stable, and what its properties are. We will see that the structure of the model must be altered in two different ways, depending on whether the material is hard or soft. Models of magnetic phenomena that are based on physical principles will be more accurate and have fewer parameters. Therefore, the appropriate modification will be made on the basis of the physical principles that underlie the process. This will result in a stable Preisach function that will no longer have the congruency property. It will still have the deletion property, a subject for the next chapter.

4.2 HARD MATERIALS

We will view hard materials as consisting of particles or grains that can support only a single or at most a few domains. Each domain will be assumed to be a single hysteron with two stable states. Since an isolated magnetic particle has a symmetrical hysteresis loop, particle interaction is thought to be the cause for the asymmetry of hysterons throughout the Preisach plane. The source of asymmetry is particle interaction of two types: exchange and magnetostatic. Exchange can at best be due to nearest neighbors. If it is very strong, two hysterons act as a single hysteron, thereby reducing the number of independent particles. This has noise considerations as discussed in Chapter 7. If it is weaker, it is either negligible or can be included in the magnetostatic interaction. In any case, exchange will not be considered by itself.

To characterize media, an accurate model for this interaction must be developed [1]. The local field that each particle experiences is the sum of the external applied field, the demagnetizing field, and the interaction field. The interaction field, that is, the local field in the absence of an applied field, fluctuates from particle to particle. A positive interaction field increases both the positive and negative values of the critical field, H_k. This makes the hysteresis loop, as viewed by an external field, appear to be asymmetrical.

Thus, we will view each hysteron as having a position on the Preisach plane that is determined by its interaction-free critical field and the interaction field that it sees. The interaction field at a particle will vary as the medium's magnetization changes, so the question of whether the Preisach function has any significance at all may well be asked. There are two aspects to this question. First we must ask whether the function is statistically stable, that is, whether at any region of the plane the hysteron density is constant. Then we must ask whether all the hysterons in that region have the same magnetization. The answer to the first question, addressed in this chapter, will result in the elimination of the congruency property. The answer to the second question, which will result in the elimination of the deletion property, is discussed in Chapter 5.

Consider an ensemble of randomly dispersed particles with moment $\mathbf{m_j}$. It is noted that if the medium is perfectly aligned, then $\mathbf{m_j}$ takes on only the values $\mathbf{1_x}$ and $-\mathbf{1_x}$, where $\mathbf{1_x}$ is the alignment axis. For more general alignments, in this chapter we will consider only a scalar model; that is, we will assume that the applied field is in a given direction and that we are interested only in the magnetization in that direction. The interaction field at a particle may be decomposed into a component along the applied field, which adds to it directly, and a component perpendicular to the applied field, which changes the critical field. If the Stoner–Wohlfarth model is applicable, the largest critical field is a little over twice the smallest one. Thus, for a particle making roughly a 20° angle with respect to the applied field, the maximum change in critical field is about ±33%. In the following analysis, we will neglect this variation. Later, we will consider the more general case.

In the scalar case, we will compute the component of the interaction field $\mathbf{H}_i$ seen by a particle. This interaction field in general is given by

$$\mathbf{H}_i = \sum_{i \neq j} \mathbf{m}_j \cdot \mathbf{T}_{ij}, \tag{4.1}$$

where $\mathbf{T}_{ij}$ is the interaction field tensor between the ith and the jth particle, and $\mathbf{m}_j$ is the moment of the jth particle. We will assume that the magnetization of each hysteron is in the x direction, and the only component of the interaction field is in the x direction. This is consistent with the idea that we are developing a scalar model. The relaxation of this condition will be discussed later in connection with vector models.

The tensor $\mathbf{T}_{ij}$ is given by

$$\mathbf{T}_{ij} = \nabla_i \nabla_j \frac{1}{4\pi r_{ij}}, \tag{4.2}$$

where the subscripts on the ∇'s indicate differentiation with respect to those coordinates. Thus, $\mathbf{T}_{ij}$ is independent of the values of the magnetization of the hysteron. For a magnetic medium that consists of a large number of randomly dispersed particles, $\mathbf{T}$ is a random variable and under certain conditions is independent and identically distributed. In particular for perfectly aligned media, (4.1) can be written as

$$\mathbf{H}_i = \frac{M_S}{F}\left(\sum_{\mathbf{m}_j=\mathbf{1}_x} V_j\mathbf{T}_{ij} - \sum_{\mathbf{m}_j=-\mathbf{1}_x} V_j\mathbf{T}_{ij}\right), \tag{4.3}$$

where F is the fraction of the volume taken up by the magnetic material whose saturation magnetization is M_S, and therefore, M_S/F is the saturation magnetization of the hysteron. Then, the *central limit theorem* applies to each of these sums, and thus the interaction field distribution in such media is expected to be Gaussian. We will make the assumption that the interaction field is Gaussian and is completely defined by two numbers: its mean and its variance.

If all subsets of $\mathbf{T}_{ij}$ are also independent of the $\boldsymbol{m}_i$, then the standard deviation is constant, and the expectation value of the interaction field is given by

$$\overline{\mathbf{H}_i} = \frac{M_S \cdot \overline{\sum_{i\neq j} Q_j V_j \mathbf{T}_{ij}}}{F} = \alpha m M_S. \tag{4.4}$$

Thus, the expectation value of the interaction field is directly proportional to the total magnetization, that is, the sum of the irreversible component and the reversible component. We will call the constant of proportionality the *moving constant*, α.

The method of the Lorentz cavity can be used to calculate α. In this method, a typical particle is replaced by an empty cavity, and the local field, due to all the other particles, is computed at this location [2]. The average value of the local field is computed by replacing all the other particles by a continuum whose average magnetization is the same as that of the particles. It is then seen that $\langle\mathbf{T}_{ij}\rangle$ is equal to the negative of the demagnetization tensor of the cavity. Thus for well-aligned highly acicular particles both $\langle\mathbf{T}_{ij}\rangle$, and thus α, are very small; however, αM_S may be substantial.

It is important not to confuse this correction for the local field with that for the demagnetizing field. Not only do the two corrections usually have opposite signs, but the local field correction is a material property and depends only on the magnetization in the immediate area of the calculation, while the demagnetizing field is a device property and depends on the entire magnetization as well as the shape of the material. However, they would be indistinguishable in an ellipsoidal sample that is uniformly magnetized. In fact, one could be used to balance out the other to simplify the identification process by using an appropriately shaped sample, as was done in [3].

To compute the standard deviation of the interaction field, one could use

$$\sigma_i^2 = \overline{(H_i - \bar{H}_i)^2} = \overline{H_i^2} - \bar{H}_i^2. \tag{4.5}$$

The computation of the standard deviation requires knowledge of the correlation between interaction tensors and can be performed only in certain special cases. We will assume that σ_i^2 is constant for most of the following calculations. The possibility that the standard deviation varies with the magnetization is considered in the variable-variance model [4], and is discussed in Section 4.4. A constant variance is apparently appropriate for longitudinal media. For particulate media, the moving constant was shown to be equal to the average of the x component of the demagnetization factors of the particles.

The variation of the critical fields of the hysterons is determined by the size, shape, orientation, and composition of the particles or grains that constitute the medium. We will assume that this distribution is log normal, since the critical fields must be positive. If the standard deviation of these fields is relatively small compared with their mean, it is possible to approximate the log-normal distribution by the normal distribution. This is usually the case for hard materials.

Thus, the Preisach function is given by

$$p(H_k, H_i) = \frac{1}{2\pi\sigma_i\sigma_k}\exp\left\{-\frac{1}{2}\left[\frac{(H_k - \bar{h}_k)^2}{\sigma_C^2} + \frac{(H_i + \alpha M)^2}{\sigma_i^2}\right]\right\}. \tag{4.6}$$

This is a Gaussian distribution whose peak moves with the magnetization of the medium, hence, this is called the *moving model.* When a field is applied to the medium, the term αM must be added to the effect of the field.

It is convenient to describe this distribution in the *operative plane*, which we will denote as the $h_i\, h_k$-*plane*, where the operative variables are defined by

$$h_i = H_i + \alpha M \quad \text{and} \quad h_k = H_k - \bar{h}_k. \tag{4.7}$$

In this representation, the Preisach function appears to be stable, and its peak is at the origin. Figure 4.1 is a block diagram of this moving model. The box "reversible field component computer" can contain any of the models for the reversible component of the magnetization as discussed in Chapter 3. For example, if it computes a magnetization-dependent reversible field, this reduces to the

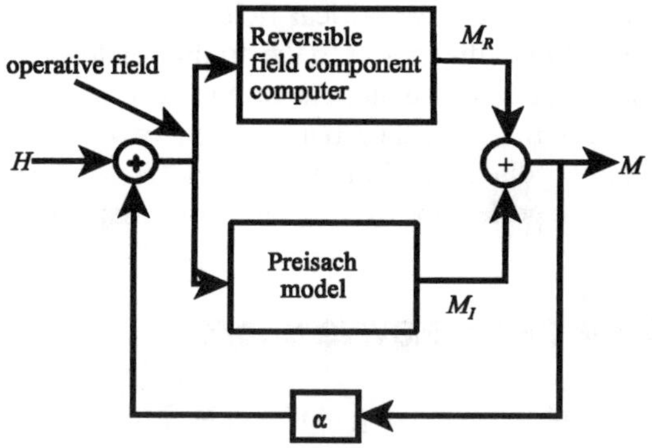

Figure 4.1 Block diagram of the moving model.

magnetization-dependent DOK model [5], but if it computes a state-dependent reversible field, then it reduces to the state-dependent CMH model [6].

The effect of the positive feedback due to the moving constant is to increase the slope of the hysteresis loop. Thus, for the same distribution, the measured switching field standard deviation σ_{meas} decreases as α increases, although the real value of σ stays the same. To see this, we first note that the slope of the hysteresis loop, the rate of change in magnetization with respect to the applied field, may be written

$$\frac{dM}{dH} = \frac{dM}{dh}\frac{dh}{dH}. \tag{4.8}$$

Then since h is $H + \alpha M$, we have

$$\frac{dh}{dH} = 1 + \alpha\frac{dM}{dH}. \tag{4.9}$$

Therefore,

$$\frac{dM}{dH} = \frac{dM/dh}{1-\alpha dM/dh}, \tag{4.10}$$

where dM/dh would be the slope of the major loop if there were no positive feedback. Since both the slope of the major loop and α are positive,

$$\frac{dM}{dH} = \begin{cases} > \dfrac{dM}{dh} & \text{if } \alpha\dfrac{dM}{dh} < 1 \\ < 0 & \text{otherwise.} \end{cases} \tag{4.11}$$

It can be shown that the same is true for the remanence loop.

For a Gaussian Preisach model, the irreversible magnetization is symmetrical with respect to the mean operative critical field. To find the relationship of the irreversible magnetization with respect to the applied field, we subtract αM from the operative field. If we take equal increments and decrements in the applied field from the mean critical field $\bar{h}_k$, the change in the reversible magnetization is smaller for the increment due to saturation. Thus, $|M_i(H - \bar{h}_k)|$ is smaller than $|M_i(H + \bar{h}_k)|$. Furthermore, the peak slope of the irreversible magnetization occurs above $\bar{h}_k$.

4.3 IDENTIFICATION OF THE MOVING MODEL

The identification of the seven moving model parameters involves identifying the four classical Preisach model parameters, $\bar{h}_k$, σ_k, σ_i , and M_S, the two reversible field parameters S and ξ, which are model-dependent, and the moving parameter α. We have already discussed techniques for identifying the first six parameters. Several methods have been proposed for the identification of the moving parameter. Since there is no best method for all values of parameters, we will now discuss two methods: the *symmetry method* and the *method of tails*. Both methods involve choosing the parameter to obtain the best fit to a desired curve.

4.3.1 The Symmetry Method

The symmetry method utilizes two facts. The first fact is that for a Gaussian Preisach function, the remanent major loop has odd symmetry about the remanent operative coercivity. This coercivity, for a single quadrant medium, is equal to the mean critical field $\bar{h}_k$. The second fact is that the moving model is a classical Preisach model when its input is the operative field. In particular, if the Preisach function is Gaussian, the major remanent loop is an error function. Thus, if $M_{\text{rem}}(H)$ is the remanent major loop,

$$M_{\text{rem}}(H + \alpha M - \bar{h}_k) = -M_{\text{rem}}[-(H + \alpha M - \bar{h}_k)]. \tag{4.12}$$

Unless the squareness is 1, a real material does not have this property because the reversible component of the magnetization is nonlinear. Thus, $M_{\text{rem}}(H - \bar{h}_k)$ is not the negative of $-M_{\text{rem}}(-H - \bar{h}_k)$ and hence the major loop is not symmetrical about the coercivity. This was the principal cause of second harmonic distortion when *dc*-bias recording was used. The problem with simply finding the value of α that minimizes this difference is that we do not know what $\bar{h}_k$ is. It differs from the remanent coercivity by αM, and although M_i is zero there, M_{rem} is not.

The method, therefore, involves measuring the major hysteresis loop and then finding the value of α that minimizes the following integral:

$$I(\alpha) = \int_0^{\bar{H}_k} \left\{ M_r(H + \alpha M - \bar{h}_k) + M_r[-(H + \alpha M - \bar{h}_k)] \right\}^2 dH. \tag{4.13}$$

We can replace $\bar{h}_k$ by the remanent coercivity, H_{RC} plus $\alpha M(\bar{h}_k)$, which is $H_{RC}+\alpha M_r(\bar{h}_k)$, in evaluating this integral. Differentiating this with respect to α and finding the value that makes it equal to zero is another method of measuring α. Thus,

$$\frac{1}{2}\frac{dI(\alpha)}{d\alpha} = \int_0^{M_r}\left[M_r(H+\alpha M-\bar{h}_k)+M_r(-H-\alpha M+\bar{h}_k)\right]dM = 0. \tag{4.14}$$

The first method involves finding a minimum, while the second method involves finding a zero crossing. Thus, the second method is more sensitive. When the second method was attempted on several recording media, a precise α was found that reduces the value of I by many orders of magnitude and was limited only by experimental error [7].

The identification of the parameters in the CMH model must be performed in a particular order. The technique we are now presenting applies to single-quadrant media. The first step is to measure the major hysteresis loop $M_J(H)$ and the major remanent loop $M_{\text{rem}}(H)$ as a function of the applied field H. The first two parameters identified are the saturation magnetization M_S and the squareness S, which are defined by

$$M_S = M_J(\infty) \quad \text{and} \quad S = M_{\text{rem}}(\infty)/M_S. \tag{4.15}$$

The operative field h is given by

$$h = H + \alpha M. \tag{4.16}$$

We expect the remanence to be symmetric function of the operative field, with respect to the remanent operative coercivity. We use this criterion as a method of choosing the correct value for α. This method requires a good starting set of parameters, which we will now obtain.

As shown in Fig. 4.2, we define the coercivity, H_C to be the field at which the major loop magnetization is zero, and we define the remanent coercivity, H_{RC}, to be the field at which the major remanent loop magnetization is zero. On the ascending major loop, let

$$M_{RC} = M_J(H_{RC}), \tag{4.17}$$

and

$$M_C = M_{\text{rem}}(H_C). \tag{4.18}$$

We note that M_C is usually negative. Let H_1 be the field at which the remanent magnetization is the positive quantity, M_{RC}. That is,

$$M_{\text{rem}}(H_1) = -M_C. \tag{4.19}$$

We will also define

$$M_1 = M_J(H_1). \tag{4.20}$$

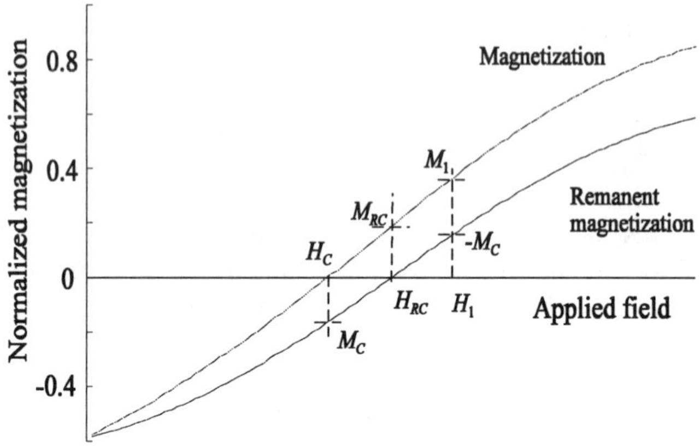

Figure 4.2 Definition of terms.

Usually M_1 and M_{RC} are positive quantities. Then a first approximation for the moving coefficient is given by

$$\alpha = \frac{H_1 + H_C - 2H_{RC}}{2M_{RC} - M_1}. \tag{4.21}$$

The effect of this choice of α is to make the curve symmetrical, as a function of the operative field, at three points. This approximation can be quite poor, since in (4.21) $2M_{RC}$ is only slightly larger than M_1. Thus, it is recommended to use (4.14) if possible.

In this case, the average operative critical field is given by

$$\bar{h}_k = H_{RC} + \alpha M_{RC}. \tag{4.22}$$

For a Gaussian Preisach function, the remanent curve as a function of the operative field should be an error function given by

$$m_{\text{rem}}(h) = \frac{M_{\text{rem}}(h)}{M_S} = S\text{erf}\left(\frac{h - \bar{h}_k}{\sigma}\right) = S\text{erf}\left(\frac{H - H_{RC} + \alpha[M(H) - M_{RC}]}{\sigma}\right). \tag{4.23}$$

Since the error function of 0.25 is 0.2763, we can define H_2 as the field that makes the normalized remanence 28% of the saturation. Then, an approximation of σ is given by

$$\sigma = 4\{H_2 - H_{RC} + \alpha[M(H_2) - M_{RC}]\}. \tag{4.24}$$

This value can be quite rough, since the approximation for σ given by (4.24) was determined by only a few measurements. A better approximation of the standard deviation of the critical field σ and the moving parameter α for a Gaussian Preisach function, is obtained by fitting $M_{\mathrm{rem}}(h)$ to an error function. Then, (4.21) and (4.24) could be used as a starting point for a two-variable search algorithm.

To describe completely the irreversible component of the magnetization we divide the standard deviation of the switching field into the standard deviation of the critical field σ_k and the standard deviation of the interaction field σ_i. To perform this separation, we saturate the material in the positive direction, apply a field $-\bar{h}_k$, followed by a field $\bar{h}_k$, and then measure the magnetization M_k, where $\bar{h}_k$ is given by (4.22). If we define

$$r = \frac{M_k}{SM_S}, \tag{4.25}$$

then

$$\sigma_i = \sigma\cos(\tan^{-1}r) \quad \text{and} \quad \sigma_k = \sigma\sin(\tan^{-1}r). \tag{4.26}$$

We note that with this definition,

$$\sigma^2 = \sigma_i^2 + \sigma_k^2. \tag{4.27}$$

To identify the reversible component of the magnetization, we need the susceptibility at zero field, χ_0. We now have the seven parameters of the CMH model: M_S, S, $\bar{h}_k$, α, σ_i, σ_k, and χ_0.

For three-quadrant media, the identification is a bit more complicated because the remanence is not the same as the irreversible component of the magnetization. Thus, S is not given by (4.15); rather, we use a modification of (4.23) with S as an additional parameter to fit. This modification is

$$\frac{M_i}{M_S} = S\,\mathrm{erf}\left(\frac{h - \bar{h}_k}{\sigma}\right), \tag{4.28}$$

where M_i is the remanence plus the change in magnetization due to apparent reversal M_{AR}. Hence

$$M_i = M_{\mathrm{rem}} + M_{\mathrm{AR}}. \tag{4.29}$$

Since M_i is not observable, we have to compute it from the observable M_{rem}, as discussed in Chapter 3. Then, in computing χ_0, we have to subtract the zero-field slope of (4.28) from the remanence susceptibility.

4.3.2 The Method of Tails

The *method of tails* subjects the material to an anhysteretic magnetizing process. If we use the classical Preisach function, the resulting remanence magnetization can be computed in closed form and is given by

$$M_r = M_S \operatorname{erf}(H_{dc}). \tag{4.30}$$

The anhysteretic susceptibility, the derivative of (4.4) with respect to H_{dc} is also Gaussian.

To test the validity of this analysis, a sample of γ-Fe_2O_3 magnetic recording tape was measured. This sample was chosen to avoid the problems of apparent reversible magnetization [8]. The data points in Fig. 4.3 show the result of a normalized anhysteretic susceptibility measurement using a vibrating sample magnetometer (VSM), and the solid line is a Gaussian fit to these data [9]. It is seen that this fit is good for only small values of H_{dc}. For large values of H_{dc}, the measurements appear to decrease more slowly to zero than the Gaussian approximation. We will now see that the moving model corrects this apparent error.

Since the Preisach function is constant only in the operative plane, this curve should have been plotted as a function of the operative field. In an experiment, it was assumed that for small values of H_{dc} the magnetization is a linear function of the applied field, the operative field is directly proportional to the applied field, and no distortion in the Gaussian occurs. As the value of H_{dc} increases, however, the magnetization saturates and the operative field does not increase as quickly with the applied field as before. This results in a tail that goes to zero much more slowly with respect to the applied field than the Gaussian.

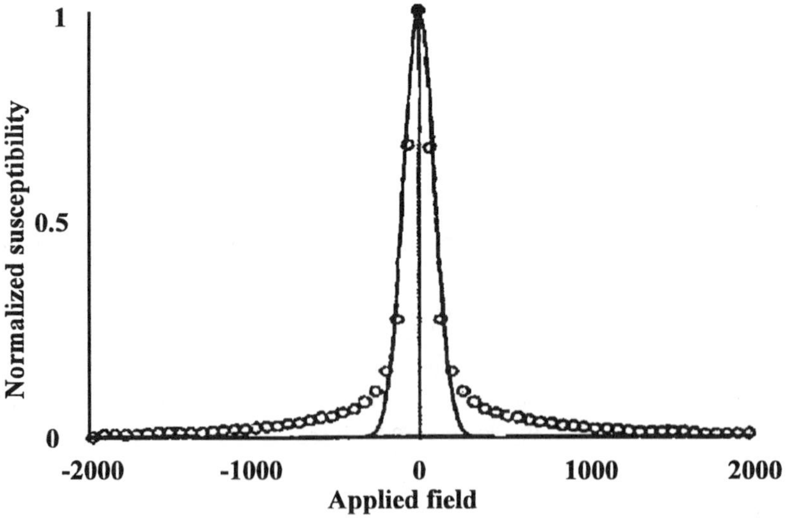

Figure 4.3 Anhysteretic susceptibility of γ-Fe_2O_3 recording tape.

The correct way to obtain a Gaussian anhysteretic susceptibility is to apply an anhysteretic sequence of operative fields rather than ordinary fields. This requires a priori knowledge of α and the simultaneous measurement of the magnetization to compute the correct current operative field. Then the field is slowly increased until the desired operative field is reached without overshooting. This is possible to do using a computer-controlled VSM.

To correct these measurements, the susceptibility was plotted as a function of the remanent operative field. That is, the remanent magnetization was multiplied by α and added to H_{dc} to approximate the operative field. The data points in Fig. 4.4 indicate the susceptibility measurements as a function of the operative field. The solid line is a fit of this curve with a Gaussian function of the operative field. Since the curve is symmetrical about H_{DC} equal to 0, only the values for positive values of H_{DC} are shown. This Gaussian function had a standard deviation of 2200 versus 130 for the Gaussian in Fig. 4.4, since it has to compensate for the moving constant. It is seen that this fit still accurately describes the region where the values of H_{dc} are small, but now it also fits the tail correctly.

This correction assumes that when an anhysteretic field process is applied to the material, the operative field is also anhysteretic. If the applied field is anhysteretic, the operative field is only approximately anhysteretic; thus, the line dividing the operative Preisach plane is only approximately straight and does not quite have the correct slope. Furthermore, we have neglected the reversible component of the magnetization in the magnetizing process. These corrections are exact only at remanence; however, the errors tend to cancel each other out. In order

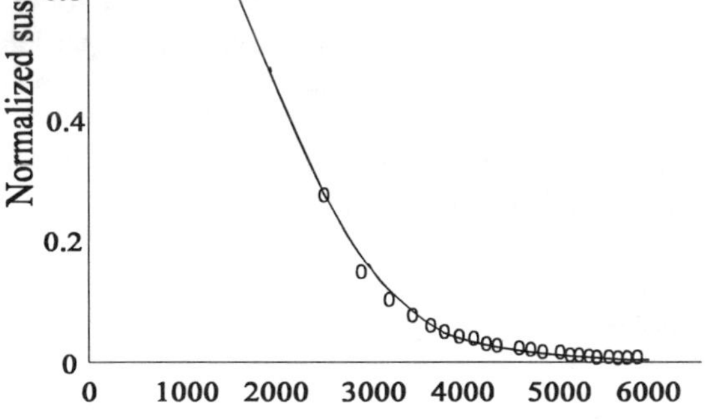

Figure 4.4 Gaussian fit to the operative susceptibility.

to be certain about the nature of the interaction field, operative field measurements must be performed.

The approximation of the interaction field distribution of the Preisach function by a Gaussian function is based on a theoretical application of the Central Limit Theorem. Furthermore, the mean value of this Gaussian is directly proportional to the magnetization, and the standard deviation appears to be constant during anhysteretic magnetizing processes. Measurements on a sample of γ-Fe_2O_3 recording tape show that this is an excellent approximation if plotted as a function of the operative field.

4.4 THE VARIABLE-VARIANCE MODEL

In longitudinal recording media, a hysteron is surrounded by particles in all directions; hence, even if they are all magnetized in the same direction, there can be a considerable variation in the local field. In some thin-film perpendicular media, the material is uniformly one grain thick, with all the grains well aligned and similarly situated. When the material is saturated, it is expected that the variation in the interaction field will be small; however, in the demagnetized state, half the grains are magnetized in one direction and half in the other direction. Thus, each grain may be surrounded by a substantially different field, leading to a much higher standard deviation in the interaction field.

In work using an artificial magneto-optic medium [10], it has been shown that for perpendicular media, the standard deviation of the interaction field is magnetization dependent. For this medium [11], the variance dependence on magnetization appears to be linear and is smaller in the demagnetized state. For the Co-Cr medium, it is seen that the variance also depends on the magnetization, but in this case the demagnetized state has the smaller variance. The basic difference is that the artificial medium was very dilute, and therefore, one considers only the few nearest neighbors. For this material, in the demagnetized state a particle can see a wide variety of configurations, but in the saturated state the configuration is very uniform. The Co-Cr medium, on the other hand, is very dense and the field at any given particle is the sum of many neighbors. Surface roughness can cause this interaction field to vary considerably from particle to particle, and its standard deviation is linear with the magnetization. At zero magnetization, there is a minimum variance in the distribution.

In these perpendicular media, the demagnetizing field shears the hysteresis loop. It has been shown that this shearing depends on the demagnetizing factor, the thickness, and the magnetic properties of the film [12]. In the full description of the model, these demagnetizing effects will be combined into an effective moving parameter. Paraphrasing our earlier work [4], we now describe this medium in terms of a model in which the variance varies with the magnetization. In addition to the identification of the other parameters, one now must also specify the variation of the standard deviation of the switching field σ. The identification is simplified in this medium, since the reversible component appears to be negligible.

This removes one of the parameters from consideration. Although a general identification strategy is not possible, since the Preisach function is not limited to the fourth quadrant, it is still possible to model accurately the major loop for this medium.

We now illustrate the effect of the remaining parameters and show how they can be identified, explaining the nature of the major loop only. The Preisach function is described in the operative plane, since it is statistically stable there. In the variable-variance model, the operative interaction field h_i is obtained by dividing $H + \alpha M$, the operative interaction field, by the standard deviation in the interaction field, σ_i. If σ_i is a constant, it simply acts as a change of scale. In the variable-variance model, σ_i is not a constant, but a function of the magnetization. It is assumed that the Preisach function is Gaussian in both the interaction field and the critical field in the operative plane. If we define the operative critical field h_i by H_k/σ_k, then the Preisach function is given by

$$P(h_i,h_k) = A \exp\left[-\frac{h_i^2 + (h_k - \bar{h}_k)^2}{2}\right], \tag{4.31}$$

where A is a suitable constant and $\bar{h}_k$ is the operative remanent coercivity. Since the critical field of a particle is determined by its physical properties only and not the magnetic state of the system, it is reasonable to expect that σ_k is constant. Therefore, we will assume that only σ_i varies and that it is a function only of M. We note that in obtaining the major loop only the switching field variance is required, which is given by

$$\sigma^2 = \sigma_i^2 + \sigma_k^2. \tag{4.32}$$

The particular variation that we will assume for σ_i as a function of the magnetization is given by

$$\sigma_i = \sigma_{i0}(1 - \nu|M|^k), \tag{4.33}$$

where ν and k are suitably chosen constants. If either ν or k is zero, the variance is constant, and the model reduces to the ordinary state-dependent model. Since the measurements indicated that the reversible component is negligible, we will assume that it is zero, causing the model to reduce to the simple moving model. The block diagram for this model is shown in Fig. 4.5. The term "modified Preisach transducer" indicates that it includes reversible components as discussed previously.

The major hysteresis loop of a magnetically uniaxial, *rf* sputtered Co-Cr film with 23% Cr, deposited on a silicon substrate, was measured using a computer-controlled vibrating sample magnetometer [4]. To ensure proper nucleation and to obtain a nearly perpendicular anisotropy over the entire thickness of the film, the film was deposited on a germanium seed layer over an SiO_2 layer on the substrate. The major loops along the film plane and perpendicular to the film plane are shown in Fig. 4.6. It is seen that the in-plane hysteresis curve is much narrower and almost

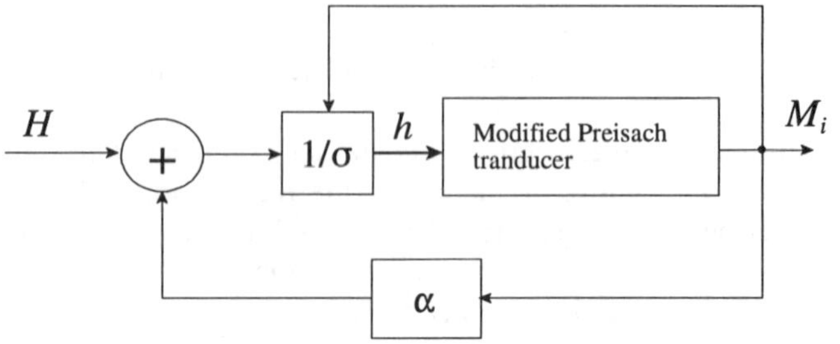

Figure 4.5 Block diagram of the variable-variance moving model for media with no reversible component.

completely reversible. On the other hand, the normal hysteresis curve has a negligible reversible component, and the knee of the ascending major loop occurs for negative values of the applied field. Furthermore, it is observed that the ascending major branch is asymmetrical, and its susceptibility below the coercivity is greater that above the coercivity. The measured saturation magnetization is 292 kA/m, and the squareness of the loop is 0.28.

Since the knee of the magnetization of the ascending major loop of this material occurs when the applied field is still negative, and since we are assuming that there is no reversible component, this implies that this increase is due to *apparent reversible* magnetization [13]. Apparent reversible magnetization is due to "particles" or *hysterons*, both of whose switching fields are negative. These hysterons are described by points in the third quadrant of the Preisach plane. The presence of such particles introduces an error in the routine identification of the irreversible component of the magnetization by measuring the corresponding

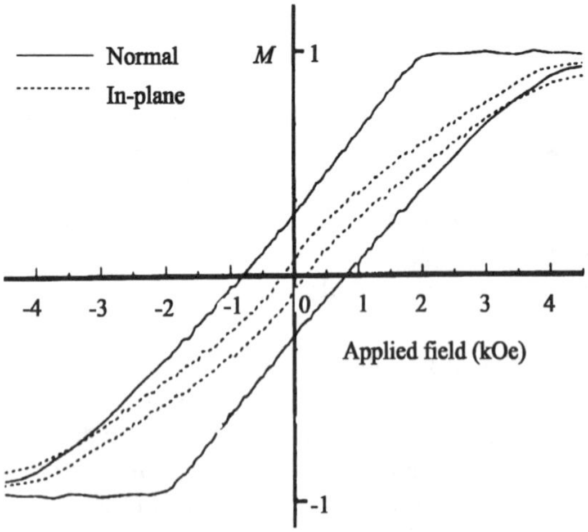

Figure 4.6 Normalized major hysteresis loops of Co-Cr perpendicular media, measured normal to and in the film plane.

remanence. In fact, the location of the knee of the ascending major loop indicates whether the Preisach function spills over into the third quadrant, and by symmetry whether it spills into the first quadrant.

The major M-H loop can be computed easily for this model, since the magnetization is an error function of the operative field h. The applied field is then computed by

$$H = \sigma h - \alpha M . \tag{4.34}$$

Thus, we can compute both M and H parametrically as a function h. The technique unfortunately does not work this simply for minor loops.

The computed ascending branch of the major loop, assuming that the variance is constant ($k = 0$), is shown in Fig. 4.7. It is seen that as the moving parameter α becomes more negative, the slope of the branch decreases and the knee moves to the left. Thus, one effect of the negative α is to push the Preisach function outside the fourth quadrant as the medium becomes magnetized. In all these cases, however, the curves maintain their point symmetry about the coercivity, a property of the state-dependent model when viewed in the operative plane only.

If k is not zero, the symmetry about the coercivity is lost, as shown in Fig. 4.8. The location of the knee continues to move to the left, and the slope of the curve continues to decrease as α becomes more negative. In this case, since k is 1, the susceptibility is greater if the applied field is an amount ΔH greater than the coercivity, than if it is an amount ΔH less than the coercivity. As α increases, the slope can even become negative.

To illustrate the effect of the exponent k, the curves in Fig. 4.9 were computed. The case of $k = 0$ is the case in which the variance is constant. In the case of $k = 1$, the variance in the interaction field varies linearly with the magnetization. It is seen that the larger the value for k, the more distortion in the symmetry about the

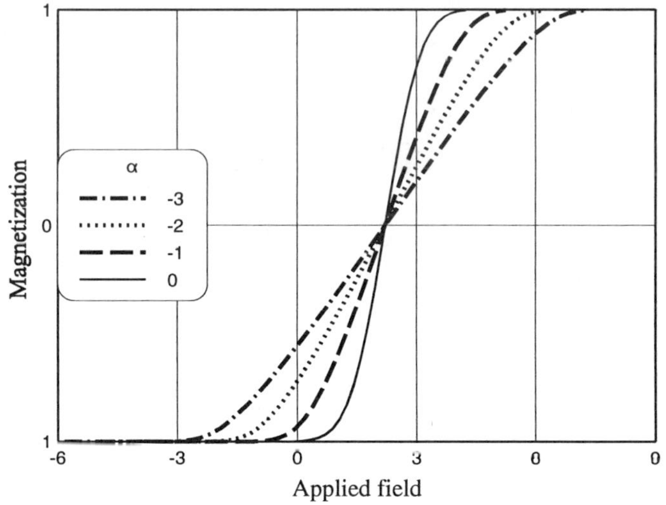

Figure 4.7 Effect of moving parameter, α, if the variance is constant ($k = 0$).

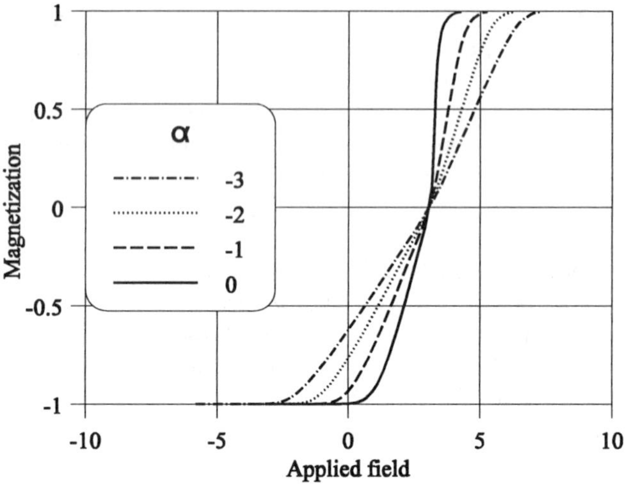

Figure 4.8 The ascending branch of the major hysteresis loop for different α's when k is 1.

coercivity. Although for nonzero values of k the coercivity is larger than the case of $k = 0$, it is not a strong function of k.

For $k = 1$, as ν is varied, we see in Fig. 4.10 that the type of asymmetry changes. In particular, for positive values of ν, the second derivative of the magnetization at the coercivity is positive, while for negative values of ν, the second derivative of the magnetization at the coercivity is negative. These observations are useful in fitting the measured curves with this model.

The identification of the parameters in the model has not been solved in

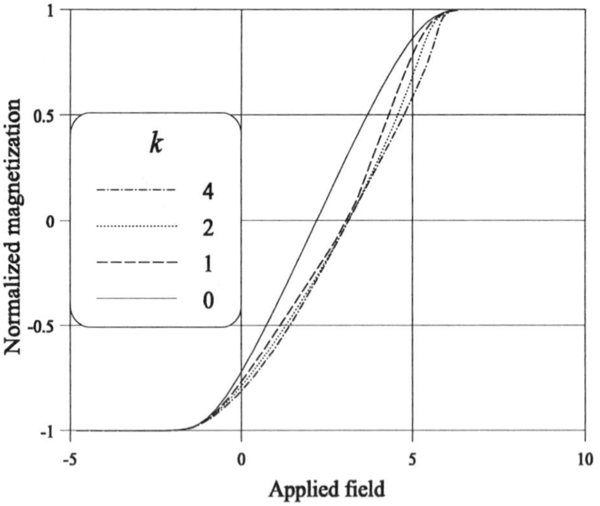

Figure 4.9 Computed ascending major branch for different values of the exponent k.

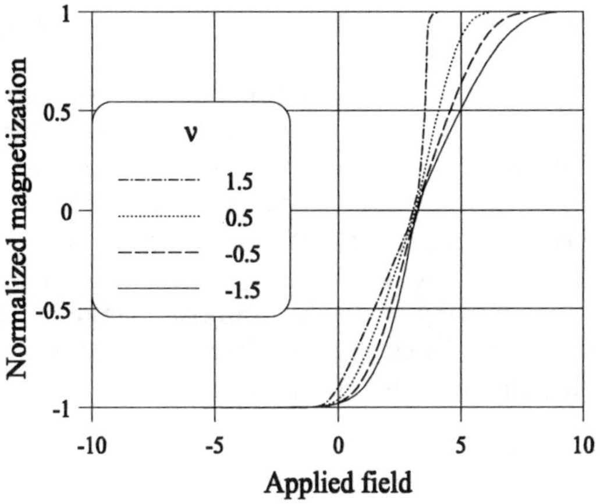

Figure 4.10 Effect of ν on the computed ascending branch of the major hysteresis loop.

general; however, for this particular medium a good fit of the major loop was obtained with the following data: $\overline{h}_k = 6$, $\sigma_i = 100$ Oe, $\sigma_k = 50$ Oe, $k = 2$, $\nu = -1.4$, and $\alpha = -2.7$. Since σ is 112 Oe, when M is zero $\sigma\bar{h}_k$ is 672 Oe. The negative ν indicates that the variance is larger when the medium is saturated. The resulting simulation is shown by the solid curve in Fig. 4.11. It is seen that the agreement is very good between the results of the model, indicated by the solid line, and the measured values, indicated by the data points. This is all the more remarkable,

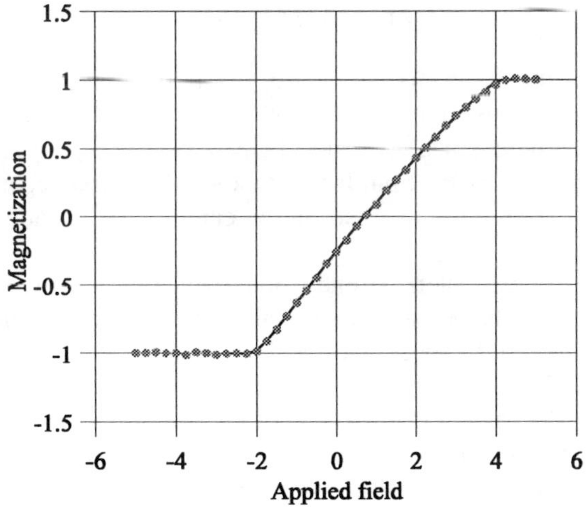

Figure 4.11 Comparison of measurements (dots) of the ascending major loop of the Co-Cr sample with the variable-variance model (solid line).

because the Preisach function in the operative plane is a very smooth and symmetrical Gaussian curve, whereas, the observed susceptibility is very asymmetrical.

The state-dependent model is a seven-parameter model that can adequately describe longitudinal media, for which it can be shown that the variance is approximately constant. This is not true in the case of thin-film Co-Cr perpendicular media, which consist essentially of a single layer of particles. Thus, the particles are surrounded by other particles in basically two dimensions, rather than three. Therefore, the interaction between any pair of particles is always negative. The resulting moving parameter, unlike longitudinal media, is negative. This parameter includes the demagnetizing effect of the shape of the medium, since on this scale it is difficult to distinguish between the two.

4.5 SOFT MATERIALS

The behavior of soft materials is different as a result of the inherent difference in the nature of the hysterons. For hard materials the hysteron was a well-defined particle or grain and always switched the same way. For soft materials, the magnetization changes by the domain wall motion from one pinning configuration to the next. In general, these configurations are not repeatable, especially when the wall moves in the opposite direction. Still, the magnetization changes in quantum jumps, and the overall effects are similar.

The main difference is that the probability of a hysteron switching in one direction is essentially independent of that switching in the opposite direction. If we then interpret the Preisach function as the probability density function that a hysteron has switching fields U and V, which are statistically independent, then they can be factored into the product of the individual switching fields; that is,

$$p(U,V) = p(U)p(V). \tag{4.35}$$

To comply with the symmetry of nature, the individual probabilities must be identical. Thus, if the distribution is Gaussian, it is described by a single standard deviation. Furthermore, since $p(U)$ must be zero if U is negative, there can be no apparent reversible magnetization. If the standard deviation becomes comparable or larger than the coercivity, the distribution cannot be Gaussian but is probably log-normal.

In the case of soft materials, one can ask whether the Preisach function is stable. In most cases, the domains are so shaped that their demagnetization factor is zero. Thus, we would expect α to be zero. On the other hand, the number of hysterons available for Barkhausen jumps depends on the magnetization. In particular, for the last hysteron to switch before saturation is reached, there is only one possibility. In the case of a demagnetized specimen, we can have a long domain wall with many possibilities for hysterons to switch. This leads to the product model [14] proposed by Kádár, which assumes that the Preisach function is the product of a function of magnetization and a function of the switching fields.

Therefore, the rate of change in the magnetization with respect to an applied field is given by

$$\frac{dm}{dH} = K(|m|)\int_0^H p(U)dU, \tag{4.36}$$

where K is a function of the magnitude of the magnetization and must be zero when $|m|$ is unity. The simplest such function is $1 - m^2$, which we use in the subsequent examples. When the material is saturated, the susceptibility is zero. Hence, the magnetization cannot be changed until a reversal has been nucleated, so that m must be incrementally reduced from unity.

In this model, a magnetization-dependent reversible magnetization can be added very easily by including an additional single-valued function in (4.36). Thus it is seen that for any choice of $\chi(H)$, the magnetization cannot exceed saturation, since $K(|m|)$ will not permit it. The simplest choice for χ is a constant.

For example, if K is $1 - m^2$ and χ is a constant, let us apply a positive field large enough to saturate the sample and reduce it to zero. When a positive field H is reapplied, all the changes in magnetization will then be reversible. In that case, (4.36) will reduce to

$$\int_{m(0)}^{m(H)} \frac{dm}{1 - m^2} = \int_0^H \chi \, dU. \tag{4.37}$$

On integrating we obtain

$$m(H) = \tanh[\chi H + \tanh^{-1} m(0)]. \tag{4.38}$$

This is a very reasonable curve for the reversible component when the irreversible component is saturated.

A block diagram for this model appears in Fig. 4.12. The main difference between the product model and the moving model is that the former uses multiplicative feedback instead of additive feedback. We will see that the product model, like the moving model, also violates the congruency property.

4.6 HENKEL PLOTS

Wohlfarth suggested that if there were no interaction between hysterons, the slope of the major remanent hysteresis loop would be twice the slope of the remanent virgin curve for any applied field. He did not specify the method of demagnetizing the material, since if there were no interaction, it wouldn't matter. This suggested a method of measuring the amount of interaction. If we let $m_V(H)$ be the virgin remanent curve and $m_J(H)$ be the major remanent curve, a plot of $m_J(H)$ as a function of $m_V(H)$, called a Henkel plot, should be a straight line from (0, –1) to (1,2). Any deviation from a straight line would then be due to interaction. An alternate method of measuring interaction would be to plot $m_J(H) - 2\,m_v(H)$, called

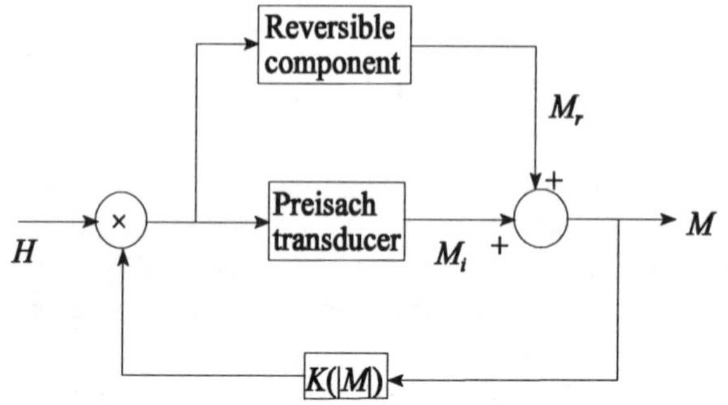

Figure 4.12 Block diagram of the product model.

Δm, as a function of the applied field. This plot would be a horizontal line through the origin if there were no interaction.

Measurements of this sort gave curves that fell both above and below the noninteracting locus. Bertotti et al. have shown [15] that for the classical Preisach model, only curves beneath that locus are predicted, but in the moving model, both types of behavior are possible. Except for the case of noninteracting particles, the virgin curve is different depending on how the material is demagnetized. In the following analysis, we will assume that *ac*-demagnetization is used to obtain the virgin curve.

From (2.27), the major remanence loop, for α equal to zero, that is, the classical Preisach model, is given by

$$\frac{dm_j}{dH} = \frac{1}{\sigma}\sqrt{\frac{2}{\pi}}\exp-\frac{(H-\bar{h}_k)^2}{2\sigma^2}. \tag{4.39}$$

By substituting $-H$ for H_1 in (2.23) we obtain the virgin magnetization curve given by

$$\frac{dm_V}{dH} = \frac{\exp\left[-(H-\bar{h}_k)^2/2\sigma^2\right]}{\sigma\sqrt{2\pi}}\left[1 - \operatorname{erf}\left(\frac{\kappa\bar{h}_k-(1-\lambda)H}{\tau}\right)\right], \tag{4.40}$$

which reduces to

$$\frac{dm_V}{dH} = \frac{\exp\left[-(H-k)^2/2\sigma^2\right]}{\sigma\sqrt{2\pi}}\left\{1 - \operatorname{erf}\left[\frac{\sigma_I(\bar{h}_k-H)}{\sigma_k}\right]\right\}. \tag{4.41}$$

Thus, we see that if there is no interaction, σ_i is zero. Then the argument of the error function is zero, and thus, the error function itself is zero. Therefore,

$$\frac{dm_V}{dH} = \frac{1}{2}\frac{dm_J}{dH}. \tag{4.42}$$

This is the Wohlfarth conjecture, which states that if there is no interaction, then

$$m_J(H) = 2m_V(H) - 1. \tag{4.43}$$

To try to quantify "interaction," researchers used plots of two types to illustrate the deviation from (4.43). In Henkel plots, m_v is plotted as a function of m_J. For noninteracting materials, this should be a straight line from (0, –1) to (1, 1). In Δm curves, m_J - $2m_V$ -1 is plotted as a function of the applied field. For noninteracting materials, this should be a horizontal line through the origin.

It is seen from (4.40) that

$$\frac{dm_V}{dH} = \begin{cases} > \dfrac{1}{2}\dfrac{dm_J}{dH} & \text{if } H < H_C \\ < \dfrac{1}{2}\dfrac{dm_J}{dH} & \text{if } H > H_C. \end{cases} \tag{4.44}$$

Thus, a Δm curve would start at the origin, have a negative slope until H_C, and have a positive slope after that until it returns to zero, as H increases.

As an illustration, Δm curves for a square loop material with k equal to zero, one, two, and four are shown in Fig. 4.13. The applied field is normalized to the coercivity, and σ is 0.3. It is seen that if there is no interaction, then K is equal to zero, and indeed a horizontal line is the result. As k increases, so does the deviation from the horizontal line. The slope of the Δm curves is negative up to the coercivity and positive after that.

When α is a positive number, positive feedback is introduced around the transducer. If there is no reversible magnetization, then

$$\frac{dm}{dH} = \frac{dm}{dh}\frac{dh}{dH} = \frac{dm}{dh}\left(1 + \alpha\frac{dm}{dH}\right). \tag{4.45}$$

Since α, dm/dH, and dm/dh are all positive, we have

$$\frac{dm}{dH} = \frac{dm/dh}{1 - \alpha\, dm/dh} > \frac{dm}{dh}. \tag{4.46}$$

Thus, the effect of the moving constant, α, is to increase the slope of the magnetization curve. Its effect on Δm, as a function of the applied field normalized to the coercivity, is illustrated in Fig. 4.14, when k is two for a square loop material. It is seen that for the classical Preisach model, α is equal to zero, and the Δm curve is always negative. When α is not equal to zero, the Δm curve can be positive as well as negative.

4.7 CONGRUENCY PROPERTY

The relationships between the *moving model* and the *product model* are explored by examining both the irreversible and the reversible susceptibility variations

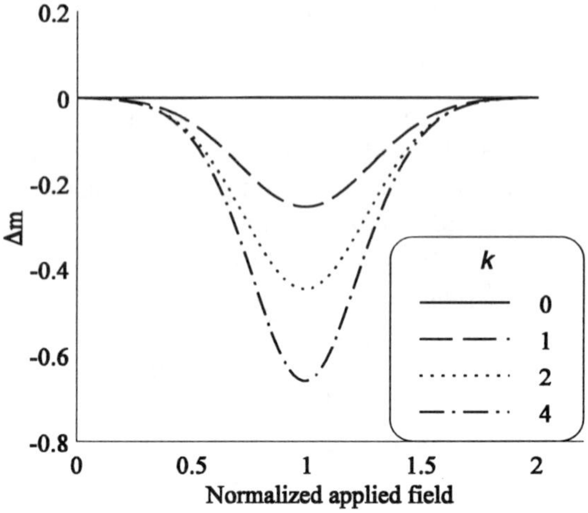

Figure 4.13 Henkel plot for classical Preisach model.

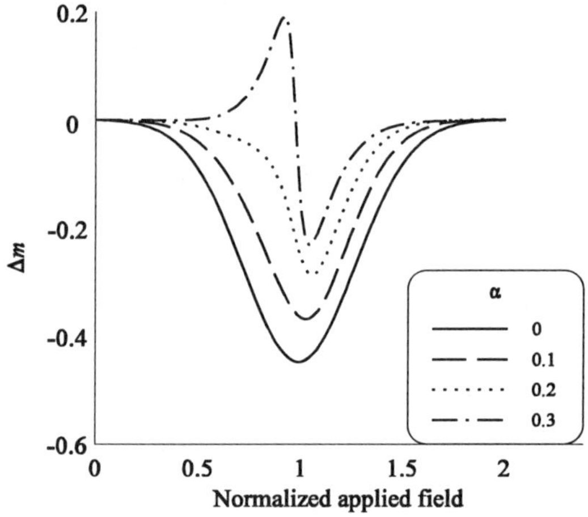

Figure 4.14 Δm curve for various moving constants.

predicted by each model [16]. It is shown that for the moving model these susceptibilities are a function of the sum of the applied field and a term proportional to the total magnetization. For the product model they are the product of a function of the total magnetization and a function of the applied field. This leads to a different variation in the height of minor loops, and thus, a means of differentiating between the models. Measurements reported elsewhere show that for particulate magnetic recording media, the moving model yields more realistic results. The reversible magnetization component of the moving model had to be modified by devising two new models, for the reversible magnetization, compatible with the moving model.

We now discuss how these properties may be used to differentiate between the various models, considering each model's unique way of circumventing the congruency property by means of an examination of the irreversible susceptibility predicted by each model. The results of experiments [17] indicate that for a particulate γ-Fe_2O_3 medium, the model most applicable appears to be the moving model; the same work gives measurements of the variations in the reversible susceptibility in the interior region of the hysteresis loop. We will discuss the properties of several models for the reversible component in order to compare them with experiment.

4.7.1 The Classical Preisach Model

The classical Preisach model computes the irreversible magnetization using

$$M_i = \int\int_{u>v} Q(u,v)\,P(u,v)\,du\,dv, \tag{4.47}$$

where $P(u,v)$ is the Preisach density function of the positive and negative switching fields. The function Q is process dependent, and for scalar irreversible models can take on only the values -1 or $+1$, depending on the sequence of applied input field extrema. The change in magnetization when the applied field is increased from H_1 to H_2, can be computed from the Everett integral:

$$E(H_1,H_2) = \int_{H_1}^{H_2} dv \int_{H_1}^{v} du\; P(u,v). \tag{4.48}$$

Therefore, we define the irreversible susceptibility for the classical Preisach model χ_{ci} as

$$\chi_{ci}(H) = \frac{E(H,H+\Delta H)}{\Delta H}, \tag{4.49}$$

where ΔH is a small increment in the applied field. This can be interpreted as the ratio of the height of an incremental minor loop to its width. It is seen from (4.49) that for the classical Preisach model, the susceptibility is a function of the applied field and the width of the minor loop. Since the susceptibility varies with the size of ΔH and is in fact zero when ΔH is zero, in all subsequent calculations, we will use the same value for ΔH.

To demonstrate the congruency property of the classical Preisach model, we point out that the susceptibility is not a function of the magnetization. This is illustrated in Fig. 4.15, which shows the variation of the susceptibility predicted by this model in the interior of the major hysteresis loop.

4.7.2 Output-Dependent Models

The effect of the moving model is to replace the applied field H in the classical Preisach model with an operative field, $h = H + \alpha M$, where α, the moving parameter, is a constant for a given medium. Thus, the irreversible susceptibility

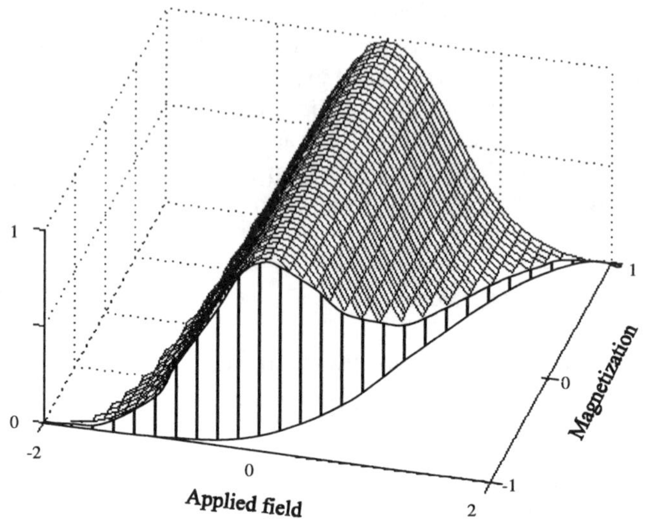

Figure 4.15 Variation in susceptibility in the classical Preisach model.

of the moving model, χ_{mi}, is a function of the magnetization as well as of the applied field and is given by

$$\chi_{mi}(H,M) = \chi'_{ci}(H + \alpha M). \tag{4.50}$$

The prime on χ'_{ci} indicates that the Preisach function P was modified to P' in the moving model.

The product model [14], on the other hand, is defined by its property of giving its susceptibility directly. If the reversible component is zero, then the irreversible susceptibility of the product model, χ_{pi}, is given by

$$\chi_{pi}(H,M) = \frac{dM_i}{dH} = K(M)\int_{-\infty}^{H} P''(u,H)du = K(M)\chi''_{ci}(H), \tag{4.51}$$

where $K(M)$ is the *noncongruency function*, $P''(u,H)$ is the *residual Preisach* function of the product model, and $\chi''_{ci}(H)$ is the modified classical irreversible susceptibility. The double prime indicates the product model modification.

From the control point of view, the moving model is a nonlinear feedback process, as shown in the block diagram of Fig. 4.12 [18]. Thus, it is necessary to solve for the magnetization iteratively. This process is computationally less efficient than the product model in which the magnetization-dependent and the field-dependent parts of the modified Preisach function are separated. On the other hand, the moving model can be directly related to physical material parameters. Therefore, it is desirable to understand the relationship between the two models. The moving model relaxes the congruency limitation of the classical Preisach

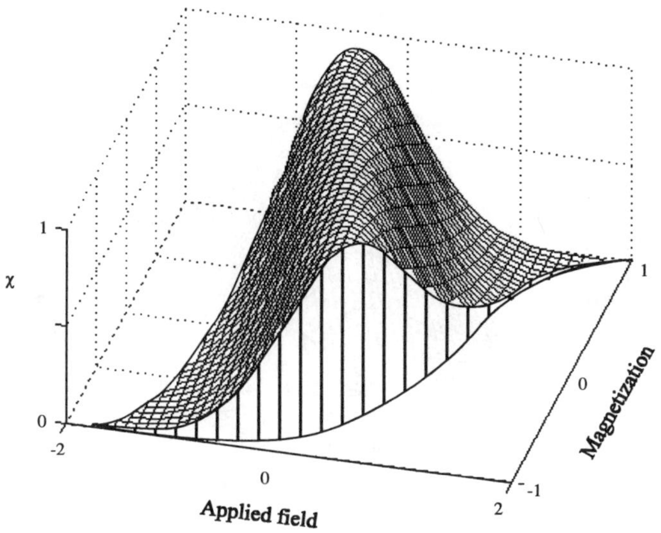

Figure 4.16 Variation in the susceptibility in the moving model.

model, replacing it with the *skew-congruency* property. Thus, minor loops connected by lines of slope $-1/\alpha$ are congruent, as illustrated in Fig. 4.16.

In addition, the product model from a control point of view is a simple Preisach transducer followed by a nonlinearity [18]. This model does not involve feedback because of the assumption that the magnetization-dependent and the field-dependent parts of the Preisach function can be separated. Thus, the identification problem is greatly simplified: $K(M)$ is obtained by measuring the variation in height of minor loops along the M axis, and the residual Preisach function $P''(u,H)$ can be used to obtain first-order reversal curve information [19]. The product model replaces the congruency limitation of the classical model by the *nonlinear congruency* property, which is equivalent to the existence of the nonlinear function $S(N)$. A plot of the variation of the susceptibility for the product model is shown in Fig. 4.17.

For the moving model, from (4.50), it is seen that

$$\chi_{mi}(H,M) = \chi_{mi}(H+\alpha M,0). \tag{4.52}$$

That is, for any line parallel to the H axis, the variation in the susceptibility is given by the variation along the M axis shifted by the amount αM. Thus, the susceptibility peak along a line parallel to the H axis will not occur on the M axis, as shown in Fig. 4.18(a). Also from (4.50), it is seen that

$$\chi_{mi}(H,M) = \chi_{mi}\left(0, M+\frac{H}{\alpha}\right). \tag{4.53}$$

That is, for any line parallel to the M axis, the variation in the susceptibility is given by the variation along the H axis shifted by the amount H/α. Thus, the

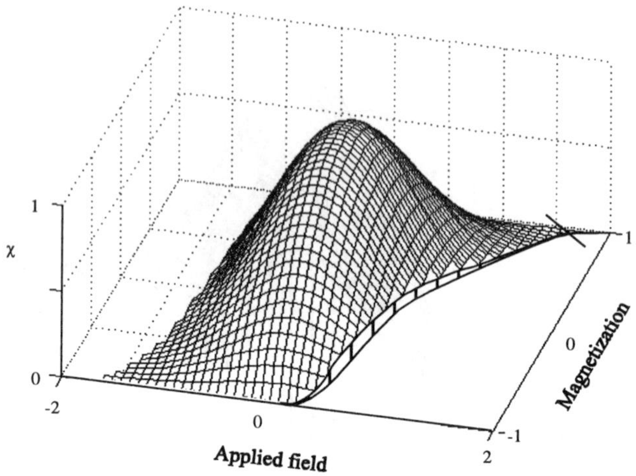

Figure 4.17 Variation of the susceptibility in the product model.

susceptibility peak along a line parallel to the M axis will not occur on the H axis, as shown in Fig. 4.18(b).

For the product model, on the other hand, from (4.52) it follows that

$$\chi_{pi}(H,M) = \chi_{pi}(H,0)\frac{K(M)}{K(0)}. \tag{4.54}$$

Thus, the variation in the susceptibility along any axis parallel to the M axis is the same. It also follows from (4.50) that

$$\chi_{pi}(H,M) = \chi_{pi}(0,M)\frac{\chi_{ci}''(H)}{\chi_{ci}''(0)}. \tag{4.55}$$

Thus, the variation in the susceptibility along any axis parallel to the H axis is also the same. Therefore, if $\chi_{ci}''(H)$ is symmetrical, then all projections of the susceptibility along any axis parallel to the H axis are symmetrical, as shown in Fig. 4.19(a). Similarly, since $K(M)$ is symmetrical [19], all projections of the susceptibility along any axis parallel to the M axis are symmetrical, as shown in Fig. 4.19(b).

4.8 DELETION PROPERTY

Some interesting properties of the Preisach model obtained in [20] will be described here. The moving model computes the irreversible component of the magnetization, M_i, using

$$M_i = \iint_{v>w} Q(w,v)P(w+\alpha M, v+\alpha M)\,dv\,dw, \tag{4.56}$$

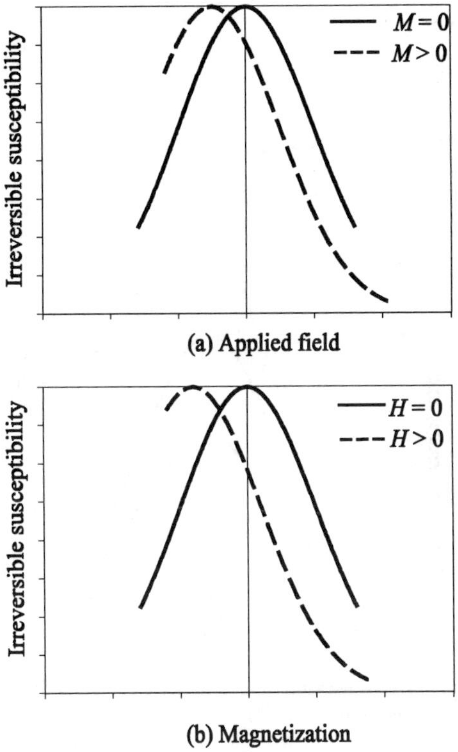

Figure 4.18 Preisach cross sections for the moving model.

where Q is a process-dependent function, which for scalar processes is either +1 or −1, P is the Preisach probability density function, which is by definition greater than zero, M is the total magnetization, and w and v are the Preisach variables (i.e., the positive and negative switching fields, respectively). The limit of integration is the entire region of the Preisach plane where $v > w$, that is, the hatched region in Fig. 4.20.

A line that consists of horizontal and vertical segments only, as shown in Fig. 4.20, is the boundary that separates the simply connected region where Q is +1 from the simply connected region where Q is −1. In the case of an anhysteretic magnetization process, it becomes in the limit of many cycles a continuous curve with a negative slope. The sequence of discontinuities in the boundary in the first Preisach variable w_k is a monotonically increasing sequence in k corresponding to the sequence of successive maxima of the input variable, while the corresponding sequence in the second variable v_k is a monotonically decreasing sequence in k corresponding to the successive minima of the input variable. In the case of the moving model, the input variable is the sum of the applied field plus the product of α and the magnetization.

We define the Everett integral by:

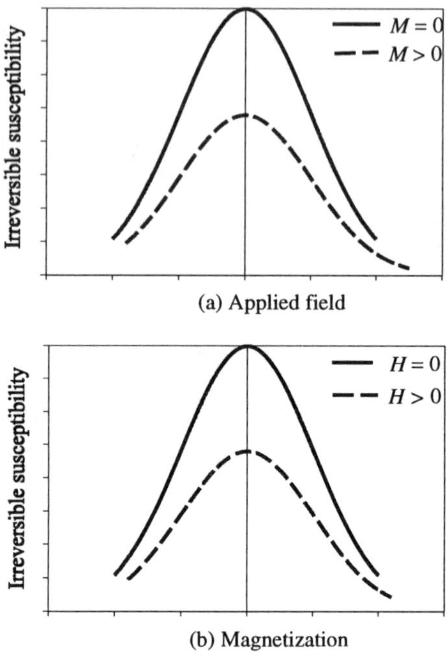

Figure 4.19 Projection of the irreversible susceptibility for the product model along an axis parallel to (a) the H axis, and (b) the M axis.

$$E(r,s) = \int_r^s \int_r^v P(w,v)\,dw\,dv. \tag{4.57}$$

If the applied field is increased from H_1 to H_2 with a corresponding increase in magnetization from M_1 to M_2, and if $H_2 + \alpha M_2$ is less than the previous maximum of $H_2 + \alpha M_2$, the change in magnetization can be computed from the Everett integral:

$$M_2 - M_1 = E(H_1 + \alpha M_1, H_2 + \alpha M_2). \tag{4.58}$$

The same formula applies if the applied field is decreased from H_1 to H_2 with a corresponding decrease in magnetization from M_1 to M_2, and if $H_2 + \alpha M_2$ is greater than the previous minima of $H_2 + \alpha M_2$. In order to compute the magnetization M_2, (4.58) must be solved implicitly.

4.8.1 Hysteresis in Intrinsically Nonhysteretic Materials

For materials that have no intrinsic hysteresis, P is a delta function in w whose line of singularity is the line w equal to $-v$. In this case, the Preisach function can be expressed as a function of a single variable, the applied field H. The magnetization

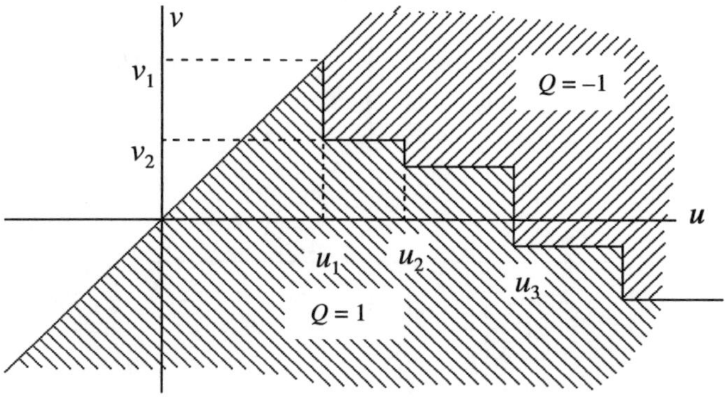

Figure 4.20 Boundary between the regions of opposites signs of Q.

is the cumulative distribution of the Preisach function, $F(H)$, and it increases monotonically.

The moving model can introduce hysteresis for these materials if α is sufficiently large. This can be seen by examining Fig. 4.21, which shows a typical plot of $F(H)$ for such a material, and how it is modified by αM to obtain the curve $F(H + \alpha M)$. It is seen that for $|H| < H_k$, there are three possible values of M. The central value is unstable, but the other two values are stable. At $H = H_k$, if one is on the lower curve, the magnetization will switch discontinuously to the upper curve, as indicated by the dashed line, leading to extrinsic hysteresis. For $|H| > H_k$, the curve is single-valued. In materials in which dF/dH is a monotonically decreasing function, the condition for hysteresis is $\alpha > 1/\chi$, where χ is the initial susceptibility; that is, χ is dF/dH at $H = 0$.

The behavior is more complex for materials in which dF/dH increases at first

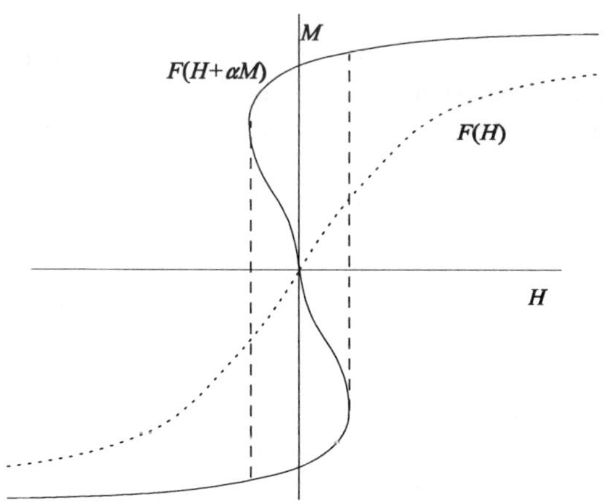

Figure 4.21 Effect of α on magnetization process.

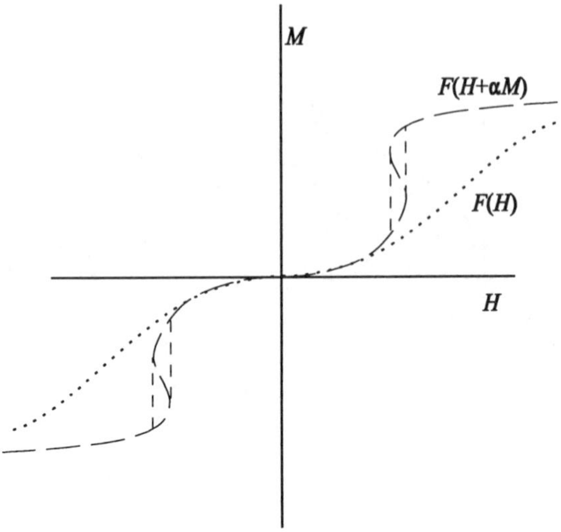

Figure 4.22 Effect of α on the magnetizing process when χ_{max} is at a positive H.

and then decreases monotonically to zero. In this case, there are again two possible states at zero H, but four discontinuous regions of operation. For small values of α, as shown in Fig. 4.22, there is no hysteresis at zero H, but there are two minor hysteresis loops symmetrically displaced from the origin. The condition for this type of hysteresis to occur is $\alpha > 1/\chi_{max}$, where χ_{max} is the maximum susceptibility. If α is increased further to a second critical value, the situation pictured in Fig. 4.23 is obtained. If one starts with a demagnetized sample, at a certain critical field a jump occurs to the major loop after which it is not possible to demagnetize the sample by any sequence of applied fields. The behavior in this case outwardly appears to have simple hysteresis.

4.8.2 Proof of The Deletion Property

According to the deletion property, the final state of magnetization is the same if a local maximum and its subsequent local minimum are deleted whenever they are followed by a larger local maximum. This sequence results in a shorter sequence and guarantees that all minor loops close. The same is true if the roles of maxima and minima are interchanged. This deletion is illustrated in Fig. 4.24 where the maximum labeled a and the subsequent minimum labeled b may be deleted from the sequence of extrema that define the magnetic state of the system. The proof of this is based on the fact that the magnetic state at point a' is the same as at point a.

The magnetic state is completely defined by the boundary line, shown in Fig. 4.20, dividing the Preisach plane into the region where Q is -1 from the region where Q is $+1$. To show that the moving model has the *deletion* property, it is necessary only to show that the same boundary configuration is attained when a minor extremum is encountered and the same applied field is returned to. That is,

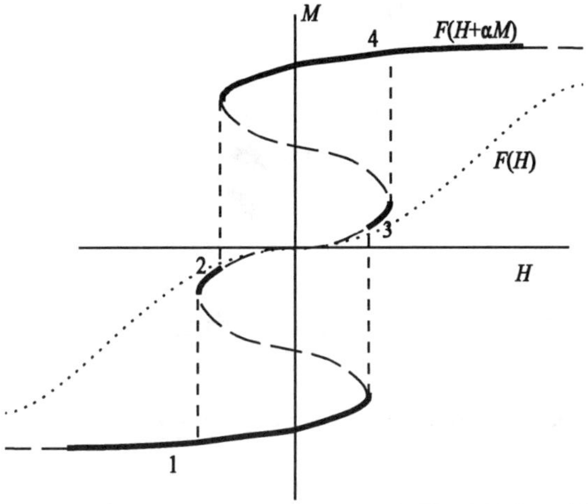

Figure 4.23 Similar magnetizing process as in Fig. 4.20, but with larger α.

the minor loop in going from H_b to H_c and back to H_b is a closed loop, as shown in Fig. 4.25.

A rigorous mathematical proof of this property is beyond our scope. We instead give a heuristic proof based on the properties of the Everett integral shown in Fig. 4.26. The Everett integral $E(r,s)$ is a monotonically increasing function of s that saturates if s is large. It is also zero when s is equal to r and has a slope, $\eta = \partial E/\partial s$, that is zero at that point. Furthermore, it is an odd function with an interchange of its arguments, so that $E(r,s)$ is equal to $-E(s,r)$. Starting from a given applied field, H_0, with a corresponding magnetization, M_0, to find the change in magnetization ΔM, it is necessary to find the solution to

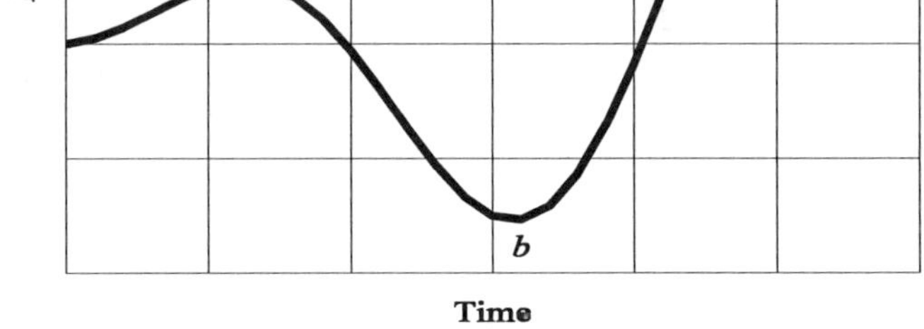

Figure 4.24 A sequence of applied fields in which extrema a and b are deleted by maximum c.

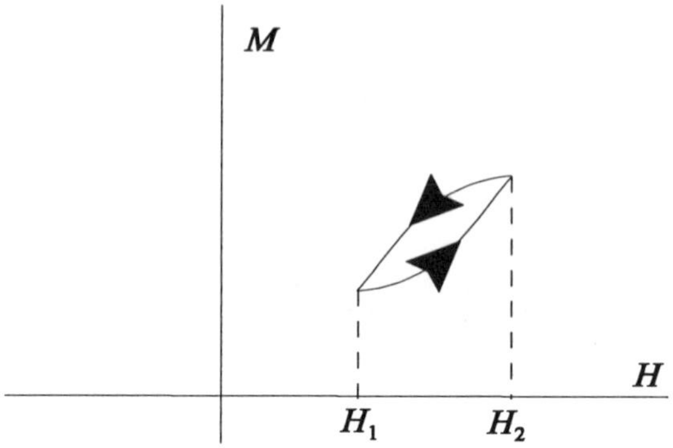

Figure 4.25 Minor loop predicted by the moving model.

$$\Delta M = E(r, r + \Delta H + \alpha \Delta M), \tag{4.59}$$

where $r = H_0 + \alpha M_0$ and ΔH is the change in applied field from H_0. The solution can be found graphically by locating the intersection of the Everett integral curve and the straight line intersecting the s–r axis at ΔH with slope $1/\alpha$. The solution is unique as long as α is less than $1/\eta_{max}$. When α is greater than $1/\eta_{max}$, then for certain fields there can be three possible solutions; however, only the lowest one is physically realizable. In that case, there may be a discontinuity in the magnetization when the applied field is increased to the point that only a single solution exists again. This is illustrated in Fig. 4.27, which shows how the moving model transfer function is constructed from the Everett integral.

The change in magnetization in going from H_b to H_c is given by $E(H_b + \alpha M_b, H_c + \alpha M_c)$. Similarly, in going from H_c to $H_{b'}$ the change in magnetization is given by $E(H_c + \alpha M_c, H_{b'} + \alpha M_{b'})$. Since the properties of the Everett integral

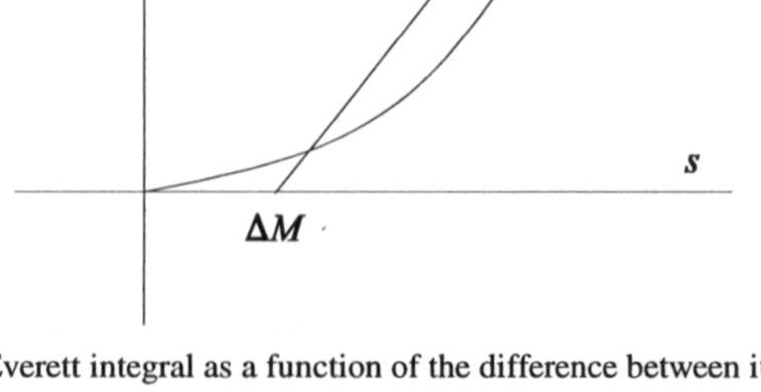

Figure 4.26 Everett integral as a function of the difference between its arguments.

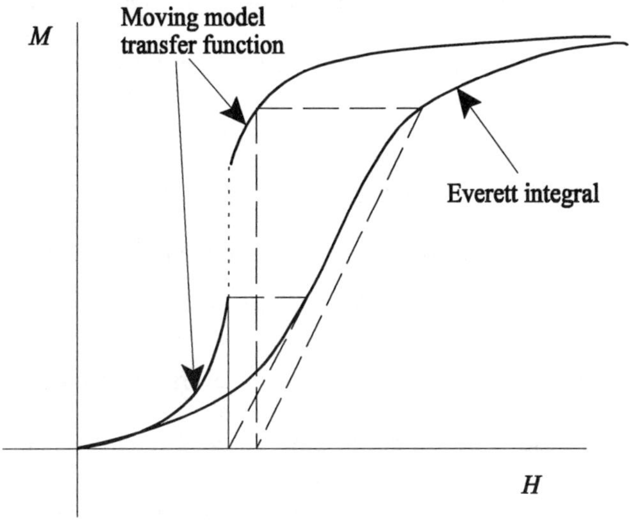

Figure 4.27 Construction of the moving model transfer function from the Everett integral.

lead to a unique solution, we must have $M_b = M_{b'}$, and thus, the minor loop is closed. Even if a reversible component of magnetization is added to the irreversible component computed by the Everett integral, the proof holds provided the reversible component is a function of the applied field and the irreversible magnetization only.

A direct consequence of the deletion property is that a process having this property cannot have accommodation, since returning to the same applied field must produce the same final state. Thus, to be able to reproduce accommodation, a further modification of the model must be made. Elsewhere [21] we have suggested such a modification. The next chapter shows that accommodation models do not have the *deletion* property.

4.9 CONCLUSIONS

We now summarize the results of the last three chapters. Four models have been presented for the irreversible magnetization: the classical Preisach model, the moving model, the product model, and the variable-variance model. In addition, we presented three models for the locally reversible magnetization: the state-independent model, the magnetization-dependent model, and the state-dependent model. Each of these models has its own characteristic, and we may take any irreversible magnetization model and add it to any locally reversible magnetization model and obtain a new model. These models can be used to describe any material with varying degrees of accuracy. If it is not important to characterize all the effects that the more accurate models were devised to do, choose the model that is sufficiently accurate for the desired application but also is most efficient computationally.

The concept of an operative field permits one to use the formulation of the classical Preisach model with either the moving model or the variable-variance model. This in effect distorts the field axis so that irreversible susceptibility is no longer symmetrical about its peak, $\bar{h}_k$. Furthermore, the peak no longer occurs at the remanent coercivity, H_{RC}, but to the left of it by the amount $\alpha M(\bar{h}_k)$. Since the irreversible component of the magnetization is zero at $\bar{h}_k$, the total magnetization $M(\bar{h}_k)$ is due purely to the locally reversible magnetization. Thus,

$$H_{RC} = \bar{h}_k - \alpha M_r(\bar{h}_k). \tag{4.60}$$

Furthermore, if we are using state-dependent reversible magnetization, $M_r(\bar{h}_k)$ is not uniquely defined unless one knows the history of the magnetizing process.

Although these models affect different portions of the magnetizing curve, and some of them remove the congruency property, they all possess the deletion property. In Chapters 2–4, we have concentrated on hysterons that are uniquely set by the applied field, ignoring the hysterons that are not supposed to be affected by it. In the next chapter, when we examine the behavior of the latter hysterons, we will find that their effect is to cause minor loops to drift. This in turn serves to remove the deletion property from the resulting model.

REFERENCES

[1] E. Della Torre and F. Vajda, "Effect of apparent reversibility on parameter estimation," *IEEE Trans. Magn.*, **MAG-33**, March 1997, pp. 1085–1092.

[2] E. Della Torre, "Effect of interaction on the magnetization of single domain particles," *IEEE Trans. Audio Electroacoust.*, **AE-14**, June 1966, pp. 86–93.

[3] E. Della Torre, "Measurements of interaction in an assembly of gamma-iron oxide particles," *J. Appl. Phys.*, **36**, February 1965, pp. 518–522.

[4] E. Della Torre, F. Vajda, M. Pardavi-Horvath, and C. J. Lodder, "Application of the variable variance model to Co-Cr perpendicular recording media," *J. Magn. Soc. Japan*, **18**, suppl. S1, 1994, pp. 117–120.

[5] E. Della Torre, J. Oti, and G. Kádár, "Preisach modeling and reversible magnetization," *IEEE Trans. Magn*, **MAG-26**, November 1990, pp. 3052–3058.

[6] F. Vajda and E. Della Torre, "Characteristics of magnetic media models," *IEEE Trans. Magn.*, **MAG-28,** September 1992, pp. 2611–2613.

[7] F. Vajda and E. Della Torre, "Measurements of output-dependent Preisach function," *IEEE Trans. Magn.*, **MAG-27**, November 1991, pp. 4757–4762.

[8] E. Della Torre and F. Vajda, "Parameter identification of the complete-moving hysteresis model using major loop data," *IEEE Trans Magn.*, **MAG-30**, November 1994, pp. 4987–5000.

[9] E. Della Torre and F. Vajda, "Computation and measurement of the

interaction field distribution in recording media," *J. Appl. Phys.*, **81**(8), April 1997, pp. 3815–3817.

[10] M. Pardavi-Horvath and G. Vertesy, "Measurement of switching properties of a regular 2-D array of Preisach particles," *IEEE Trans. Magn.*, **MAG-30**, January 1994, pp. 124–127.

[11] F. Vajda, E. Della Torre, M. Pardavi-Horvath and G. Vertesy, "A variable variance Preisach model," *IEEE Trans. Magn.*, **MAG-29**, November 1993, pp. 3793–3795.

[12] G. J. Gerritsma, M. T. H. C. W. Stam, J. C. Lodder, and Th. J. A. Popma, "Initial slope of the hysteresis curve," *J. Phys. Colloq*, C8, S12, **49**, December 1988, pp. 1997–1998.

[13] O. Benda, "To the question of the reversible processes in the Preisach model," *Electrotech. Cas.*, **42**, 1991, pp. 186–191.

[14] G. Kádár, "On the Preisach function of ferromagnetic hysteresis," *J. Appl. Phys.*, **61**, 1987, 4013–4015.

[15] V. Basso, M. Lo Bue, and G. Bertotti, "Interpretation of hysteresis curves and Henkel plots by the Preisach model," *J. Appl. Phys.*, **75**(10), May 1994, pp. 5677–5682.

[16] F. Vajda and E. Della Torre, "Minor loops in magnetization-dependent Preisach models," *IEEE Trans. Magn.*, **MAG-28**, March 1992, pp. 1245–1248.

[17] F. Vajda and E. Della Torre, "Measurements of output-dependent Preisach function (Invited)," *IEEE Trans. Magn.*, **MAG-27**, November 1991, pp. 4757–4762.

[18] E. Della Torre, "Existence of magnetization-dependent Preisach models," *IEEE Trans. Magn.*, **MAG-27**, July 1991, pp. 3697–3699.

[19] G. Kádár and E. Della Torre, "Hysteresis Modeling I: Noncongruency," *IEEE Trans. Magn.*, **MAG-23**, September 1987, pp. 2820–2822.

[20] M. Brokate and E. Della Torre, "The wiping-out property of the moving model," *IEEE Trans. Magn.*, **MAG-27**, September 1991, pp. 3811–3814.

[21] E. Della Torre and G. Kádár, "Hysteresis Modeling II: Accommodation," *IEEE Trans. Magn.*, **MAG-23**, September 1987, pp. 2823–2825.

CHAPTER 5

AFTEREFFECT AND ACCOMMODATION

5.1 INTRODUCTION

This chapter treats two further corrections to Preisach modeling: aftereffect and accommodation. Due to these effects minor loops do not in general close on themselves, so both corrections remove the deletion property of the Preisach model. They do this in different ways: one is time dependent and the other is rate independent. Both usually involve small drifts of magnetization with time, so they are easily confused with each other in many cases.

Aftereffect changes the magnetization as a function of time and is mainly due to thermal effects. A magnetization state is relatively stable if it is surrounded by an energy barrier that is sufficiently high; however, no matter how high that barrier is, the magnetization will eventually revert to the ground state. The higher the barrier, the longer before reversion to the ground state is completed. In the next section, when we discuss the relationship between the height of the barrier and the length of time needed to revert to the ground state, we will see that changing the physical size of the hysteron can change that time from a few minutes to many centuries.

Accommodation, on the other hand, is rate independent and is a direct result of the hysteretic many-body interpretation of the Preisach model. The drift in magnetization occurs only when the magnetization is cycled, and this drift is a function not of time but of the number of cycles that have elapsed. If one cycles the magnetization at a constant rate, the drift will appear to be a function of time. Both effects are interpreted here in terms of the Preisach model. The resulting modifications of the model generally agree with observations.

5.2 AFTEREFFECT

When a magnetic material is subject to a step function in the applied field, its magnetization will change very quickly to a new value and then slowly drift to a final value. The time constant associated with the first change in magnetization is of the order of nanoseconds, while the second is of the order of seconds. The first change can be computed with the models already discussed, but the latter must be computed differently and is the subject of this section. *Diffusion aftereffect* and *thermal aftereffect*, the main types identified thus far, are similar in behavior, although they have quite different causes. A history of the research in this area is given by Chikazumi [1].

A mechanism for diffusion aftereffect was first proposed by Snoek [2]. It involved the diffusion of carbon atoms in α-iron as the magnetization rotated. Since the carbon atoms occupy interstitial sites in the body-centered cubic that elongate the lattice, they reduce the magnetocrystalline anisotropy in that direction. Thus, when the magnetization is rotated, if the carbon atoms diffuse to a new position, they can lower the energy of the crystal. When a field is applied, the magnetization rotates quickly to the new position, but the diffusion is much more gradual, and the energy approaches the equilibrium value asymptotically. The time constant associated with the process is

$$\tau = \tau_0 e^{W/kT}, \tag{5.1}$$

where W is the barrier energy, and τ_0 is an appropriate constant whose dimension is time. This equation is referred to as the *Arrhenius law*. Experiments by Tomono [3] have shown that the logarithm of τ varies linearly with $1/T$. The slope that he measured for this variation corresponded to a value for W of 0.99 eV for this process.

Thermal aftereffect, on the other hand, involves the reversal of the magnetization of hysterons not the diffusion of atoms. This type of aftereffect, discovered by Preisach [4], is sometimes referred to as *magnetic viscosity* or as *trainage*. When a field is applied, all hysterons that have critical fields less than the applied field will switch very quickly; however, the remaining hysterons that have critical fields larger than the applied field would not switch at all if the temperature were absolute zero. At finite temperatures, this energy barrier can be overcome thermally. Since different hysterons have different barrier energies, they will switch at different rates. Thus, the aftereffect does not decay exponentially.

Let us assume that the rate of switching is given by (5.1), where W is now the energy barrier that must be overcome to reverse the magnetization of a hysteron. Then when a step change in the applied field occurs, the aftereffect magnetization, that is, the magnetization after the step change, is given by

$$m(t) = m(0) + f(t), \tag{5.2}$$

where

$$f(t) = \Delta m \left[1 - \int_0^{\infty} e^{-t/\tau} p_\tau(\tau) d\tau\right]. \tag{5.3}$$

In (5.3), $m(0)$ is the magnetization just after the step change, Δm is the total change in magnetization due to aftereffect, and $p_\tau(\tau)$ is the normalized probability that a hysteron will switch with time constant τ. Since all magnetizations are normalized, the maximum remanence is unity.

The proper choice of $p_\tau(\tau)$ determines the behavior of the aftereffect. Several distributions have been suggested for it. Chikazumi [1] has suggested a $1/\tau$ dependence between τ_1 and τ_2, while Aharoni [5] has suggested the Γ function dependence, also with two adjustable parameters, p and τ_0. Neither distribution has any physical basis nor any predictive power. The Preisach–Arrhenius model, on the other hand, links the phenomenon to hysteresis, suggests a distribution with only one adjustable parameter, τ_0, and can describe the variation of the aftereffect with the applied field.

Korman and Mayergoyz [6] and Bertotti [7] suggested that the dependence of the aftereffect on magnetization history could be described by the Preisach model. The following extension of their work was recently proposed [8]. If aftereffect is to be described in terms of the Preisach model, it is preferable to express the probability in terms of switching fields. To do this, let us consider the application of an operative field h to a material that has been saturated in the negative direction. For clarity, we will hold h constant throughout this process. If we are using the moving model, then since h depends on the magnetization, the applied field would have to be adjusted to keep it constant throughout the process; however, for hard materials, the decay rate is usually so small that any change in magnetization may be neglected for reasonable periods of time. For the classical model, then, α is zero, and no adjustment in the field is necessary. Hysterons whose switching fields are less than h will instantaneously be switched to positive magnetization, while the remaining ones will remain switched negatively, since they are protected by an energy barrier from switching immediately. If h is large enough, thermal energy will overcome this barrier and the material will eventually be saturated. We will discuss what is "large enough" in the next section in connection with more general magnetizing processes. The values of m_∞ and Δm for this process then are

$$m(0) = \int_{-\infty}^{h} du\ p(u) \quad \text{and} \quad \Delta m = 2\int_{h}^{\infty} du\ p(u), \tag{5.4}$$

where $p(u)$ is given by

$$p(u) = \int_{-\infty}^{\infty} p(u,v)\, dv = \frac{dm_i}{dh}. \tag{5.5}$$

The factor of 2 in Δm comes from the fact the Q in that region changes from -1 to $+1$. We extend the upper limit to infinity in the v integration, since the Preisach function is zero for v greater than u. We note that the integrations would have to be carried out in the operative plane in the case of the moving model. Furthermore, if the material is not square loop, an appropriate reversible component would have to be added. The consideration of these effects is beyond the scope of this book.

If we assume that the Preisach function is Gaussian, then

$$p(u) = \frac{1}{\sigma\sqrt{2\pi}} \exp\left[-\frac{(u-\bar{h}_k)^2}{2\sigma^2}\right], \tag{5.6}$$

where $\bar{h}_k$ is the average value of the critical field. Note that in the case of single-quadrant media, $\bar{h}_k$ is equal to the remanent coercivity. It follows that

$$m(0) = \operatorname{erf}\left(\frac{h-\bar{h}_k}{\sigma}\right), \tag{5.7}$$

and

$$\Delta m_i = \operatorname{erf}\left(\frac{h}{\sigma_i}\right) - \operatorname{erf}\left(\frac{h-\bar{h}_k}{\sigma}\right). \tag{5.8}$$

In this case, the medium will eventually become saturated as all the hysterons overcome the energy barrier. Figure 5.1 plots of Δm_i, the change in magnetization during the relaxation process, for various values of σ, when σ_k is zero and h_k is 1. It is seen that the field that maximizes Δm_i is half h_k, since this is the difference of two error functions, one centered at h_k and the other centered at zero. Since the maximum change in magnetization is limited to 2, the curve saturates at that value for small σ. The curve is symmetrical with respect to the peak only in this case, since σ_i is equal to σ when σ_k is zero. Since σ_i is always less than or equal to σ, the slope at the origin is usually steeper than at h_k, and the peak of this curve will occur at a value less than 1/2.

If we neglect the change in the energy stored in the reversible component of the magnetization, the energy required to switch a hysteron in a process described by the Preisach model is given by

$$W = \mu_0 M V(u-h), \tag{5.9}$$

where V is the average *activation volume* of the hysteron. Thus MV is the magnetic moment of the activation volume of the hysteron, h is the operative field, and $u-h$ is the additional field required to switch the hysteron. A micromagnetic study of recording media showed that it is necessary to switch only a fraction of the volume of a hysteron to cause it to reverse [9]. Observations of recording materials [10] have shown that this can range from values as small 0.2 of the hysteron's volume to the entire volume. The latter value is valid for very small particles. Thus, V is the minimum volume that has to be switched to nucleate a reversal, and MV is the

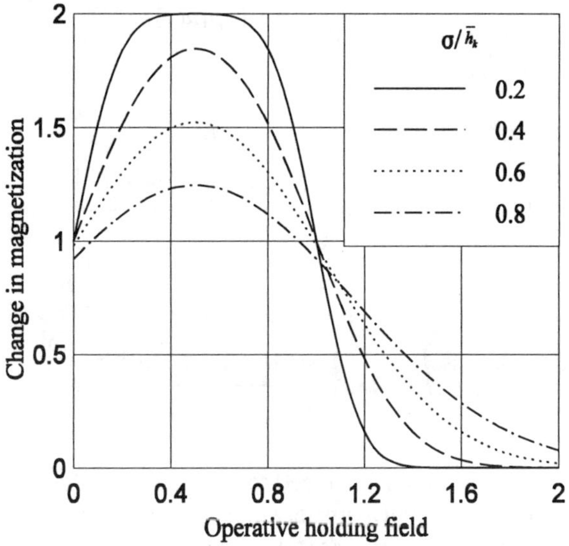

Figure 5.1 Variation in the total change in magnetization, Δm_i, with normalized holding field, $h/\bar{h}_k$, for relaxation to the ground state.

minimum moment that must be switched to reverse the entire hysteron. Then using (5.1), this hysteron would have a time constant given by

$$\tau = \tau_0 \exp\left[\frac{\mu_0 MV(u-h)}{kT}\right] \quad \text{for} \quad u>h, \tag{5.10}$$

or

$$u = h_f \ln\left(\frac{\tau}{\tau_0}\right) + h \quad \text{for} \quad \tau > \tau_0, \tag{5.11}$$

where

$$h_f = \frac{kT}{\mu_0 MV}. \tag{5.12}$$

The parameter h_f is referred to as the *fluctuation field*, and has the units of magnetic field. It is equal to the field required to make the hysteron's energy barrier equal to the thermal energy. If this factor is large compared to the switching field, the hysteron will be superparamagnetic. In the study of aftereffect, we are interested only in small values of h_f. For useful recording media h_f is small compared to the switching field of the hysteron, and therefore, its magnetization is retained for long periods of time.

We note that

$$p_\tau(\tau) = p[u(\tau)]\frac{du}{d\tau} = \frac{h_f p[u(\tau)]}{\tau}. \tag{5.13}$$

Then, using (5.6) and (5.11), $p[u(\tau)]$ is given by

$$p[u(\tau)] = \frac{1}{\sigma\sqrt{2\pi}} \exp\left\{-\frac{[h - \bar{h}_k + h_f \ln(\tau/\tau_0)]^2}{2\sigma^2}\right\}. \tag{5.14}$$

Thus,

$$f(t) = \frac{2h_f}{\sigma\sqrt{2\pi}} \int_{\tau_0}^{\infty} \frac{d\tau}{\tau} \exp\left\{-\frac{t}{\tau} - \frac{[h - \bar{h}_k + h_f \ln(\tau/\tau_0)]^2}{2\sigma^2}\right\}. \tag{5.15}$$

It is noted that the lower limit is changed to τ_0 because of the limitation imposed by (5.11). Note that $f(0)$ is approximately one as t approaches zero. If we change the variable of integration to $y = \ln(\tau/\tau_0)$, we obtain

$$f(t) = \frac{h_f}{\sigma}\sqrt{\frac{2}{\pi}} \int_0^{\infty} dy \exp\left\{-\frac{te^{-y}}{\tau_0} - \frac{[h - \bar{h}_k + h_f y]^2}{2\sigma^2}\right\}. \tag{5.16}$$

Using (5.2), we see that the magnetization as a function of time is given by

$$m(t) = 1 - \frac{\Delta m_i h_f}{\sigma}\sqrt{\frac{2}{\pi}} \int_0^{\infty} dy \exp\left\{-\frac{te^{-y}}{\tau_0} - \frac{[h - \bar{h}_k + h_f y]^2}{2\sigma^2}\right\}. \tag{5.17}$$

This shows that the amount of aftereffect is a function of the applied field.

To illustrate the time dependence of (5.17), this expression was integrated numerically and plotted on a semi-log plot in Fig. 5.2 for two values of h_f. The parameter used in the plot, which is reasonable for a recording medium with fairly large hysterons, was $\sigma = 0.6$. The value of τ_0 used in this simulation was 0.1. A field equal to the coercivity is applied so that the initial magnetization is zero. Since the hysterons that are positively magnetized will remain magnetized because of this field, and since the hysterons that are negatively magnetized will eventually also become magnetized, the magnetization will approach saturation.

It is seen that for times somewhat greater than τ_0, the magnetization increases linearly with the logarithm of the time. The effect of h_f is to change the slope of the linear portion of the aftereffect on the log-time curve. This linearity can continue for many decades, as seen from the curve when $h_f/\bar{h}_k$ is equal to 0.007; however, when the magnetization approaches about half its final value, the curve starts to deviate from the straight line, as seen from the curve when $h_f/\bar{h}_k$ is equal to 0.07. It is characteristic of this process that a small change in h_f can cause a large change

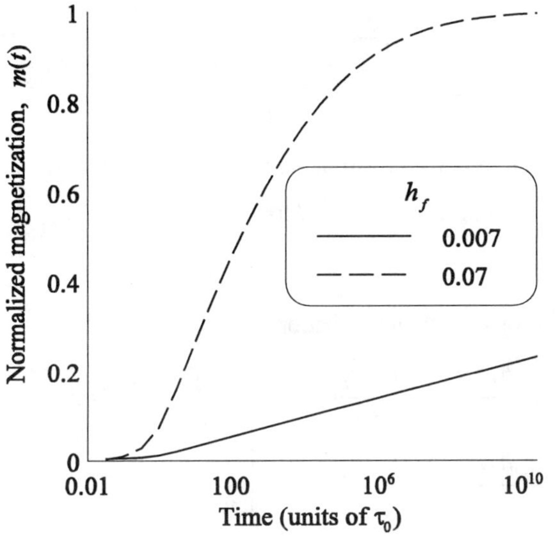

Figure 5.2 Aftereffect as a function of log time for two values of h_f.

in the behavior of the aftereffect. These results have been studied for a wide range of materials and generally agree with these conclusions [11].

It is noted that as h is increased from zero, the total range of the aftereffect decreases until when it saturates the medium, the range of the aftereffect is zero. The second effect of h is to change the slope, $\hat{S}$, of the aftereffect on the log time curve in the linear region. To evaluate the slope we first differentiate (5.17) with respect to time:

$$\frac{dm}{dt} = \frac{\Delta m_i h_f}{\tau_0 \sigma}\sqrt{\frac{2}{\pi}}\int_0^{\infty} \exp\left[-y-\frac{te^{-y}}{\tau_0}-\frac{(h-\bar{h}_k-h_f y)^2}{2\sigma^2}\right] dy. \tag{5.18}$$

We now define the *coefficient of magnetic viscosity*, $\hat{S}$, as discussed by Wohlfarth [12]:

$$\hat{S} = \frac{dM(t)}{d\log t} = SM_S\frac{dm(t)}{d\log t}. \tag{5.19}$$

This is the rate of decay of the magnetization on a logarithmic scale. It has been so defined because many materials appear to decay linearly on such a scale. We will see that for "permanent magnet" materials this is the case over a range of times that are accessible to experimenters. However, when t is very small or very large, log t diverges and the decay is no longer linear.

Since

$$\frac{d}{d\log t} = t\frac{d}{dt}, \tag{5.20}$$

we have

$$\frac{\hat{S}(t)}{SM_S} = \frac{dm(t)}{d\log t} = \frac{\Delta m_i h_f t}{\tau_0 \sigma}\sqrt{\frac{2}{\pi}}\int_0^\infty \exp\left[-y - \frac{te^{-y}}{\tau_0} - \frac{(h - \bar{h}_k + h_f y)^2}{2\sigma^2}\right] dy. \quad (5.21)$$

If h_f is small enough to permit us to neglect the term $h_f y$, this reduces to

$$\frac{\hat{S}}{SM_S} = \frac{\Delta m_i h_f t \exp[-(h - \bar{h}_k)^2/2\sigma^2]}{\tau_0 \sigma}\sqrt{\frac{2}{\pi}}\int_0^\infty \exp\left[-y - \frac{ty}{\tau_0}\right] dy. \quad (5.22)$$

If we substitute u for e^{-y}, then this becomes

$$\begin{aligned}\frac{\hat{S}}{SM_S} &= \frac{h_f t \exp[-(h - \bar{h}_k)^2/2\sigma^2]}{\tau_0 \sigma}\sqrt{\frac{2}{\pi}}\int_0^1 e^{-tu/\tau_0} du \\ &= \frac{h_f t \exp[-(h - \bar{h}_k)^2/2\sigma^2]}{\tau_0 \sigma}\sqrt{\frac{2}{\pi}}\left(1 - e^{-tu/\tau_0}\right|_0^1.\end{aligned} \quad (5.23)$$

Thus, if t is much larger than τ_0, this reduces to

$$\frac{\hat{S}}{SM_S} = \frac{\Delta m_i h_f \exp[-(h - \bar{h}_k)^2/2\sigma^2]}{\sigma}\sqrt{\frac{2}{\pi}}. \quad (5.24)$$

It is seen that the slope is independent of both t and τ_0. Furthermore, it is proportional to a Gaussian whose maximum occurs when the applied field is equal to $\bar{h}_k$ and whose standard deviation is σ. The maximum slope at h equal to $\bar{h}_k$ is $0.7979\ h_f/\sigma$.

The decay coefficient would have a maximum for $h = \bar{h}_k$ were it not for the variation in Δm_i. Since Δm_i is a decreasing function of h, the location of the peak in $\hat{S}$ must be located at a value smaller than $\bar{h}_k$. The amount of decrease in the location of the peak depends on the slope of Δm_i versus h, which is roughly inversely proportional to σ. This variation in decay coefficient with holding field has a maximum that is inversely proportional to σ, as illustrated in Fig. 5.3 for four different values of σ.

It is noted that at H_{RC} the irreversible component of the magnetization is zero, and thus the peak occurs at $H_{RC} - \alpha M_r(H_{RC})$, where $M_r(H_{RC})$ is the locally reversible (and only) component of the magnetization. It has been shown that α is a positive number less than one; thus, $\bar{h}_k$ is less than H_{RC}. There is a further correction, as discussed in Section 3.7, if the material is a three-quadrant material, that is, if σ_i is not negligible compared to the coercivity.

The irreversible susceptibility χ_i can be computed by substituting (5.6) into (5.5). Thus,

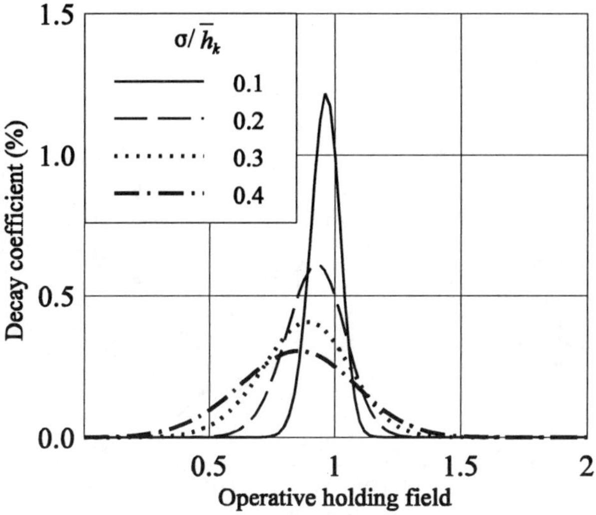

Figure 5.3 Variation in decay coefficient with holding field for various critical field distributions for negligible h_f.

$$\chi_i = p(u) = \exp\frac{[-(h-\bar{h}_k)^2/2\sigma^2]}{\sigma\sqrt{2\pi}}. \tag{5.25}$$

Then (5.22) can be written

$$\hat{S} = 2\Delta m_i h_f S M_S \chi_i. \tag{5.26}$$

This result is comparable to that obtained by Street and Woolley [13].

Figure 5.4 is a plot of (5.21) for $\sigma = 0.6$, $h_f/\bar{h}_k = 0.006$, and $\tau_0 = 1$. It is seen that for t less than $0.1\tau_0$, the slope is essentially zero. It then begins to rise, reaching about 64% of its maximum value at τ_0. By $10\tau_0$, it is essentially at its maximum, and then is essentially constant for many decades. In particular at $10^8\tau_0$ its magnitude has decreased less than 2.5% from the peak. If $h_f/\bar{h}_k$ were decreased to 0.0006, then the decrease would be less than 0.0125%.

The model accurately predicts that aftereffect is essentially linear as a function of the logarithm of time. Furthermore, the slope of this curve is a maximum around the coercivity. It assumes that hysterons in a field large enough to switch them will remain switched, but hysterons that can be in either state will on average be demagnetized. The results presented were based on the classical Preisach model for simplicity, but the corrections for motion and state-dependent reversible magnetization must be made if high accuracy is desired. The apparent reversible magnetization of three-quadrant media will affect the field that maximizes the slope of the aftereffect on the log time curve. Other effects, discussed in the coming sections, also affect these results.

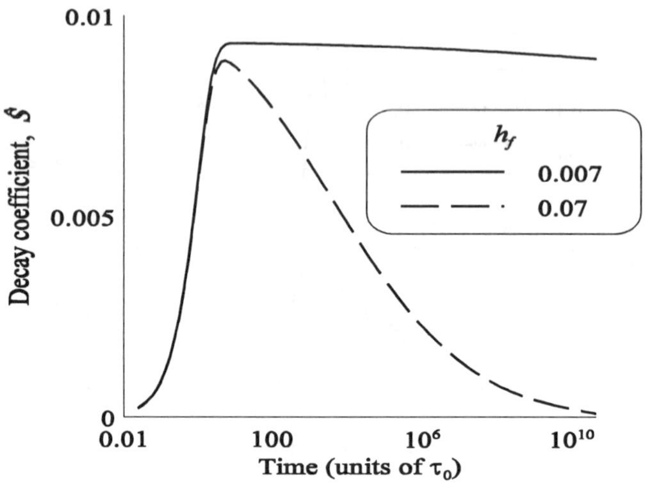

Figure 5.4 Slope of the aftereffect on a log time scale. (Note the decay coefficient is multiplied by 10 for $h_f = .007$ for direct comparison.)

This model has three parameters: σ, h_f, and τ_0. The first is the same standard deviation of the switching field distribution of the Preisach model and can be determined in the same way. The second, the fluctuation field, is defined in terms of physical quantities and can be measured, since the model predicts that the slope of the log-time variation is $1.253\ \sigma/h_f$. The last parameter, which is analogous to the mean free time between collisions in a paramagnetic gas, can be obtained by several methods. One finds that a small change in h_f will cause a small change in the $\hat{S}$; since this slope is very small, however, and since the scale is logarithmic, a small change can change the time by many orders of magnitude. Thus, doubling the size of the hysteron will cut h_f in half, but may change the time from the order of minutes to many centuries.

5.3 PREISACH INTERPRETATION OF AFTEREFFECT

Since aftereffect can be explained in terms of the Preisach model, we will now use the Preisach model to calculate aftereffect, and address the question of how to relax the restriction before (5.4) so that will h be "large enough." The preceding analysis neglects the effect of the down-switching field because if the applied field were large enough, its effect would be negligible. However, the omission leads to the wrong conclusion about the ground state magnetization. The first thing that we notice is that the aftereffect is time dependent, so a static interpretation of the Preisach diagram will not suffice. Therefore, to include time dependence, we will let the state variable Q be a function of time. We will think of Q at any point on the

Preisach plane as consisting of a fraction q_+ hysterons in the positive state and a fraction $1 - q_+$ hysterons in the negative state. Then

$$Q(t) = 2q_+(t) - 1. \qquad (5.27)$$

To derive the equations for the magnetization state of each point in the Preisach plane, let us start from negative saturation, as in Section 5.2; then $q_+(0) = 0$, and $Q = -1$, for all points in the plane. When we apply a positive field, as illustrated in Fig. 5.5, the Preisach plane is divided into three regions. Region I hysterons will be switched to the positive state and region III hysterons will be in the negative state. Hysterons in region II could be in either state, but start out in the negative state. For example, the hysterons in a small region about a point (u,v), as shown, require a field $u - h$ to switch them to the positive direction. Thus, to switch into the positive direction they must overcome an energy barrier $\mu_0 MV(u - h)$. Then their magnetization will vary exponentially with a time constant

$$\tau_+ = \tau_0 \exp\left(\frac{u-h}{h_f}\right). \qquad (5.28)$$

Similarly, any positive hysterons have to overcome an energy barrier $\mu_0 MV(h-v)$ and will do so with a time constant

$$\tau_- = \tau_0 \exp\left(\frac{h-v}{h_f}\right). \qquad (5.29)$$

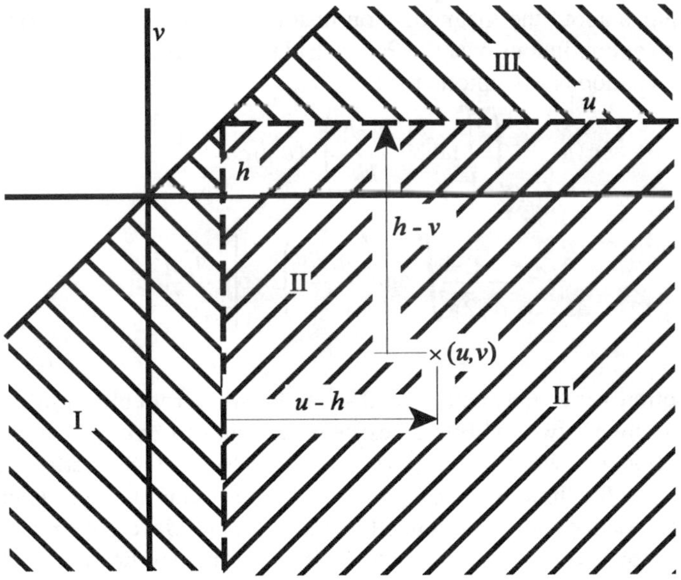

Figure 5.5 Preisach interpretation of aftereffect.

The state variable then must obey the following differential equation:

$$\frac{dq_+}{dt} = \frac{1-q_+}{\tau_-} - \frac{q_+}{\tau_+}, \tag{5.30}$$

and we can use (5.27), to convert the differential equation for the magnetization state at each point in the Preisach plane to

$$\frac{dQ}{dt} + Q\left(\frac{\tau_+ + \tau_-}{\tau_+\tau_-}\right) = \frac{\tau_+ - \tau_-}{\tau_+\tau_-}. \tag{5.31}$$

The initial condition for this differential equation is $Q = -1$. When the applied field is not constant, the τ's are functions of position; otherwise this is a first-order differential equation with constant coefficients.

For a constant applied field, it is seen that the steady-state solution of (5.31) is

$$Q(u,v)|_{t=\infty} = \frac{\tau_+ - \tau_-}{\tau_+ + \tau_-} = \tanh\left(\frac{u+v-2h}{2h_f}\right) = \tanh\left(\frac{h_i - h}{h_f}\right). \tag{5.32}$$

The time constant to reach this solution is different for each point on the plane. For example, from (5.31) at a particular point it is given by $(\tau_+ - \tau_-)/\tau_+\tau_-$, where τ_+ and τ_- are given by (5.28) and (5.29), respectively. Furthermore, points for which $u - h = h - v$ have a steady-state value of zero. Starting from a state where Q is discontinuous about the magnetization history staircase, as time progresses, the state becomes continuous over the Preisach plane. In particular, as t becomes large, the magnetization is asymptotic to

$$\begin{aligned} m_\infty(h) &= \iint_{u>v} \tanh\left(\frac{h_i - h}{h_f}\right) p(u,v)\,du\,dv \\ &= \frac{1}{\sigma\sqrt{2\pi}} \int_{-\infty}^{\infty} \tanh\left(\frac{h_i - h}{h_f}\right) \exp\left[-\frac{h_i^2}{2\sigma_i^2}\right] dh_i. \end{aligned} \tag{5.33}$$

This function varies between +1 and –1 as h varies from minus infinity to plus infinity. There are two limiting cases: when h_f goes to zero, this approaches the error function; and when σ goes to zero, this approaches the hyperbolic tangent of $(h - h_i)/h_f$. In the limit of h_f going to zero, Q approaches the sign function at $h_i = h$. This is the same result as obtained by the Korman–Mayergoyz model [6].

Aftereffect can be described by the Preisach model; however, when this is done, the process is no longer rate independent. The technique for including aftereffect in the Preisach model is to make the magnetization state a function of time. In this case, the magnetization state in any region of the Preisach plane is no longer uniform. In the next section, we will describe accommodation by a similar technique and then discuss how to compute both effects when the moving constant is not zero and when a state-dependent magnetization is added. We will see in a later section that both aftereffect and accommodation modify the deletion property of the model.

5.4 AFTEREFFECT DEPENDENCE ON MAGNETIZATION HISTORY

When a constant field is applied to a material after a complex magnetizing process, the state of each point in the Preisach plane obeys (5.31), which then may be rewritten as

$$\tau \frac{dQ}{dt} + Q = Q_\infty, \tag{5.34}$$

where

$$\tau = \frac{\tau_+ \tau_-}{\tau_+ + \tau_-} = \frac{e^{h_k/\gamma}}{2\cosh[(h_i - h)/\gamma]} \tag{5.35}$$

and

$$Q_\infty = \tanh\left(\frac{h_i - h}{\gamma}\right). \tag{5.36}$$

It is noted that Q, Q_∞, and τ are all functions of u and v. The solution to this differential equation is given by

$$Q = (Q_0 - Q_\infty)e^{-t/\tau} + Q_\infty. \tag{5.37}$$

Thus, each point in the Preisach plane must approach a different equilibrium, and each point approaches that equilibrium at a different rate.

To illustrate this effect, let us consider the first-order reversal process that starts at a large positive field and then goes to a field H_1 and finally to a field H_2, as illustrated in Fig. 5.6. The dashed line indicates the anhysteretic limit of a magnetizing process as computed by (5.33). When the applied field attains the value H_1, the resulting normalized magnetization is m_1. As the field is changed to H_2, the magnetization follows the minor loop to m_2, and finally the aftereffect causes the magnetization to drift to m_3. For example, if H_1 were the remanence coercivity, H_{RC}, m_1 would be zero. Furthermore, if H_2 were zero, m_3 would be zero. If the material were a single quadrant medium, then m_2 would also be zero; however, the magnetization will not remain at zero as the applied field is set at zero, since different points in the Preisach plane relax at different velocities. Thus, at the instant the applied field is reduced to zero, the magnetization will indeed be zero, but even though the field is maintained at zero, the magnetization will become

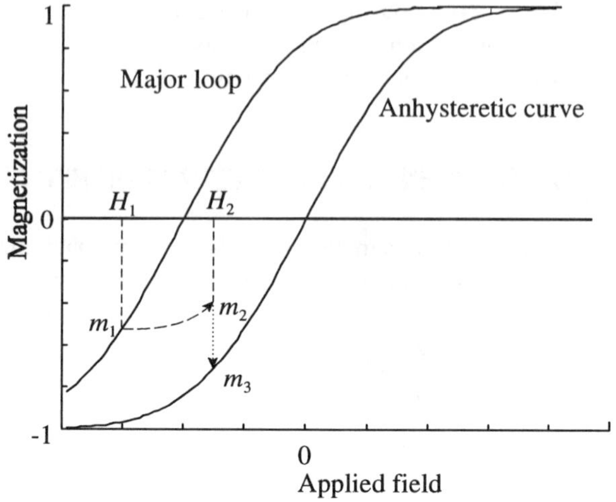

Figure 5.6 Study of the aftereffect in a first-order reversal process.

positive and then relax back to zero.

The description of this process by this model is illustrated in Fig. 5.7. When H_{RC} is applied, the region above the horizontal line at H_{RC} will become magnetized negatively, while the rest of the Preisach plane will remain magnetized positively. After the applied field is reduced to zero, the vertically hatched region will essentially revert to positive magnetization, while the horizontally hatched region will essentially become magnetized negatively. Since hysterons in the vertically hatched region have a smaller energy barrier, they will change more quickly than hysterons in the horizontally hatched region. Thus, the magnetization will first drift in a positive direction and then eventually relax back to zero. For materials that have a small fluctuation field, and thus would make good recording materials, the peak of the drift would occur after a very long time, and measurements have shown that even after days, the magnetization will continue to drift upward.

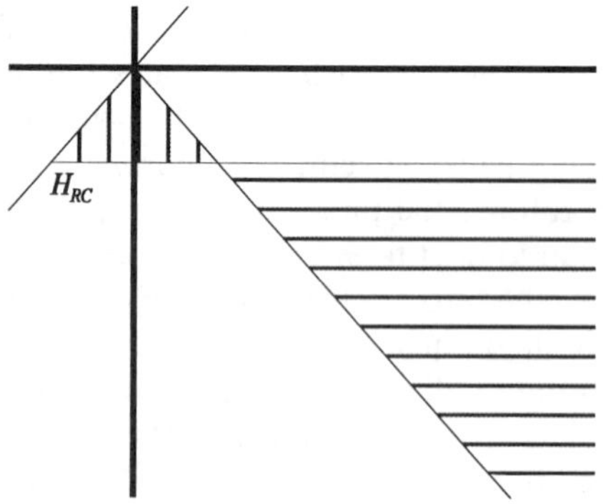

Figure 5.7 Preisach plane explanation of first-order aftereffect process.

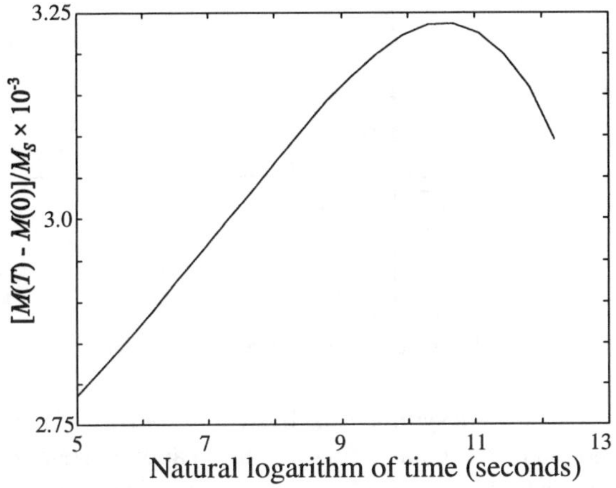

Figure 5.8 Plot of the aftereffect due to a first-order reversal process.

This process can be accelerated by not reducing the field to zero. Then the magnetization can relax to a different value but still change direction in the process. As an illustration, Fig. 5.8 shows the behavior on a log-time scale of the magnetization after a first-order reversal process in which the material started from positive saturation and then is subject to a field of –1600 Oe, which was immediately increased to –1000 Oe, and maintained at that value throughout the remainder of the measurement. The material was assumed to have a coercivity of 1140 Oe and a σ of 970 Oe, which is typical of recording media. The values used for τ_0 and h_f were 10^{-11} and 14.5 Oe, respectively. This type of behavior was observed in spring magnets by LoBue et al. [14]. Further discussion and experimental verification of these effects can be found elsewhere [15].

5.5 ACCOMMODATION

We now turn to a further statistical modification to the Preisach model to include accommodation; we will discuss the properties of such a model and the identification of its parameters. When minor hysteresis loops in magnetic media are cycled between two fields, they gradually drift toward an equilibrium loop, as shown in Fig. 5.9. This phenomenon, known as *accommodation,* requires a change in the applied field for the drift to occur. It is to be distinguished from *aftereffect,* in which drift takes place even when the applied field is held constant. Accommodation cannot be described by any of the pure Preisach models presented thus far, since they possess the *deletion property* [16]. In contrast to purely phenomenological attempts to explain this effect [17-19], we will describe a statistical interpretation of Preisach models that is not limited by the deletion property. Thus the model naturally exhibits accommodation.

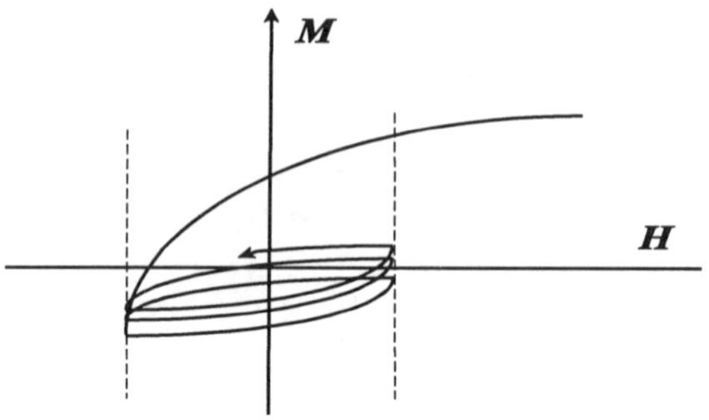

Figure 5.9 Illustration of an accommodating minor loop.

Although a hysteron has a symmetrical hysteresis loop as a function of its local field, when this loop is observed as a function of an applied field, it appears to be asymmetrical. The local field is the sum of the external applied field, which includes any long-range demagnetizing effects, and the interaction field, which is the source of the shift in the hysteron's loop. We will assume that this interaction field is Gaussian, that its standard deviation is constant, and that its mean value is proportional to the local magnetization. We will describe the Preisach functions in the *operative plane*, so that the distribution is statistically stable for most longitudinal and thick perpendicular media, which means that despite the motion of all the hysterons, the net population density at all points in the operative plane is constant.

We will now compute the irreversible magnetization component by a Preisach type model by neglecting aftereffect and using

$$M_i = SM_S \iint_{u>v} Q(u,v)\,p(u,v)\,du\,dv. \tag{5.38}$$

For nonaccommodating scalar media, the state variable is +1 in the region that is positively magnetized and −1 in the region that is negatively magnetized. When an increasing field h is applied to a magnetic material, the operative plane may be divided into three regions, as shown in Fig. 5.10. The boundary between region R_1 and R_2 is a vertical line that intersects the u axis at h. The boundary between R_1 and R_3 is the customary staircase that contains the relevant history of the magnetizing process. Region R_1 is magnetized in the positive direction by the applied field. Although R_1 is normally positively magnetized and R_3 is normally negatively magnetized, since any hysteron in these regions has a critical field greater than the applied field, any hysteron that moves into these regions can maintain its original magnetization.

When a given hysteron moves in the plane as a result of a change in its local field, it takes its magnetization with it. If in the new location it experiences a field large enough to change it, it will reverse its magnetization; however, if in the new location the hysteron experiences a field smaller than its critical field, it may not conform to the magnetization of the hysterons in that region. Thus, the magneti-

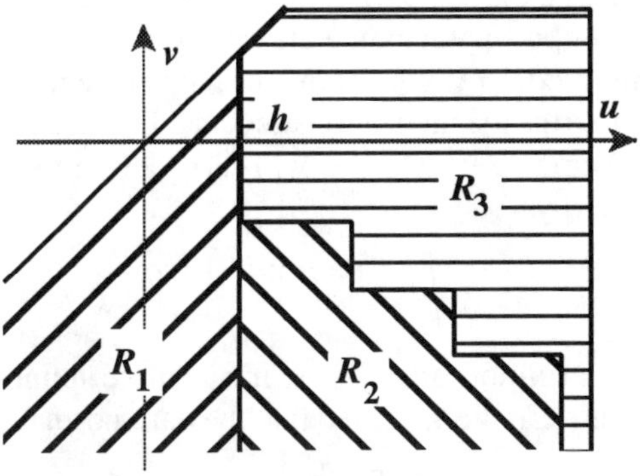

Figure 5.10 Division of the Preisach plane into three regions by an applied field.

zation of the hysteron is determined both by the region it came from and by whether in its new region it experiences a field large enough to switch it. Table 5.1 summarizes the effect of this motion in the operative Preisach plane. The column labeled "State" shows the sign of the magnetization of a hysteron that moved from the region labeled "Initial Location" to the region labeled "Final Location." On the other hand, the column labeled "Other Models" shows the sign the magnetization would have in the nonaccommodating interpretation of the Preisach model. For example, it is seen that if a hysteron originally in R_1 had ended in R_3, it would have the *"wrong"* value of magnetization.

When a hysteron has the *"wrong"* value of magnetization, it will dilute the strength of the magnetization component due to this region. The dilution will be accounted for by changing the interpretation of the state variable $Q(u,v)$ to the average of the state variables in the region, as in the aftereffect model. The change

Table 5.1 Hysteron Motion in the Preisach Plane

Initial Location	Final Location	State	Other Models
R_1	R_1	+	+
R_1	R_2	+	+
R_1	R_3	+	−
R_1	R_1	+	+
R_1	R_1	+	+
R_1	R_3	+	−
R_3	R_1	+	+
R_3	R_1	−	+
R_3	R_3	−	−

in Q normally may not be the same for all locations of the same region; however, as a first approximation, we will change all values in a given region by the same factor. Thus, in this model Q is uniform in a given region. In the following analysis, we will denote the magnetization associated with the region R_j by M_j, and the associated state variable by Q_j. We will also define the component of the Preisach function in R_j by

$$p_j = \iint_{R_j} P(u,v)\,du\,dv. \tag{5.39}$$

Note that with this definition all the p_j's are positive numbers less than one. Furthermore, with this definition, we see that they are normalized, so that

$$\sum_j p_j = 1, \tag{5.40}$$

and we can compute the irreversible component of the magnetization by means of

$$M_i = SM_S \sum_j Q_j p_j. \tag{5.41}$$

We assume that whenever the magnetization changes, the hysterons move in the Preisach plane and carry their magnetization state with them. The following analysis is restricted to longitudinal media whose distribution is both stable and Gaussian in the operative plane. Furthermore, as stated earlier, we assume that the motion of a given hysteron is not a function of its original position in the plane. In a special case not considered here, the motion of hysterons is restricted along lines for which $h_k = (u + v)/2$ is a constant. This is a reasonable approximation for thin, well-aligned films; however, more sophisticated models are possible.

The positions of all the hysterons in the Preisach plane change whenever the magnetization of even a single hysteron changes. The amount of motion of a hysteron depends on its proximity to the hysteron that switched. Since there is no motion of hysterons in the Preisach plane unless at least one hysteron switches, it is reasonable to assume that the amount of motion is proportional to the change in magnetization. Furthermore, when a hysteron moves to a new position in the Preisach plane, the probability that its magnetization is of a certain polarity is the same as the fraction of all the hysterons in the plane that have that polarity.

Consider a magnetizing process consisting of applied field extrema, which will be referred to as *events*. Let us call the fraction of hysterons replaced in an element of the Preisach plane by hysterons coming from other parts of the plane the positive *replacement factor* ξ, which is less than one. Therefore, at event $n + 1$ in a magnetizing process, the value of $Q_j^{(n+1)}$ at point j, in terms of the value of $Q_j^{(n)}$, would be given by

$$Q_j^{(n+1)} = Q_j^{(n)}(1-\xi) + \xi \bar{Q}^{(n)}, \tag{5.42}$$

where $\overline{Q}^{(n)}$ is the average $Q^{(n)}$ throughout the plane, and ξ lies between zero and one. Note that in this formulation, the magnitude of Q will always be less than one, since ξ is less than one. We expect this replacement factor to be proportional to the change in magnetization, since there will be no replacement unless the state of the system changes. Thus,

$$\xi = \frac{\beta}{SM_S} |\Delta M_i|, \tag{5.43}$$

where the proportionality constant β is a dimensionless constant for a given medium. We note that β must have a value that allows the magnitude of ξ to be less than one. Since the average value of Q at event n is given by

$$\bar{Q}^{(n)} = \frac{M_i}{SM_S}, \tag{5.44}$$

we have

$$\Delta Q_j^{(n)} = Q_j^{(n+1)} - Q_j^{(n)} = \frac{\beta |\Delta M_i^{(n)}|}{SM_S}\left(\frac{M_i^{(n)}}{SM_S} - Q_j^{(n)}\right). \tag{5.45}$$

This is the amount that Q changes in a given leg of the magnetizing process. In a continuous process, this difference equation is replaced by the differential equation

$$\frac{dQ}{dH} = \frac{\beta(M_i - SM_S Q)}{S^2 M_S^2}\left|\frac{dM_i}{dH}\right|. \tag{5.46}$$

This model, like all static Preisach hysteresis models, still is a time-independent process. We can, therefore, fully describe a magnetizing process by giving only the values of successive extrema of the applied field. The part of the process between two successive extrema will be referred to as a *leg* of the magnetizing process. Since this model does not possess the deletion property, we must consider all extrema, not only the ones that normally are undeleted.

For simplicity let us consider small hysteresis loops in a medium whose squareness is unity. We will now consider the cycling of a material with an arbitrary magnetization history between two operative fields: h_A and h_B, where $h_A > h_B$, and the difference between them is small. Then we can compute the magnetization changes by solving the differential equation (5.46) by Euler's method, with one step per leg. We will start the accommodation process from h_A letting the first leg of the process be the transition to h_B. The values of the various quantities during a given leg of the process will be denoted by a superscript containing the leg number in parentheses. Thus, the value of Q in region j at the first application of h_A will be denoted by $Q_j^{(1)}$.

When an applied field iterates between the operative fields h_A and h_B, the region labeled R_1 in Fig. 5.11 is entirely switched. In the classical Preisach model and in the moving model, the height of a minor loop between these extremities will

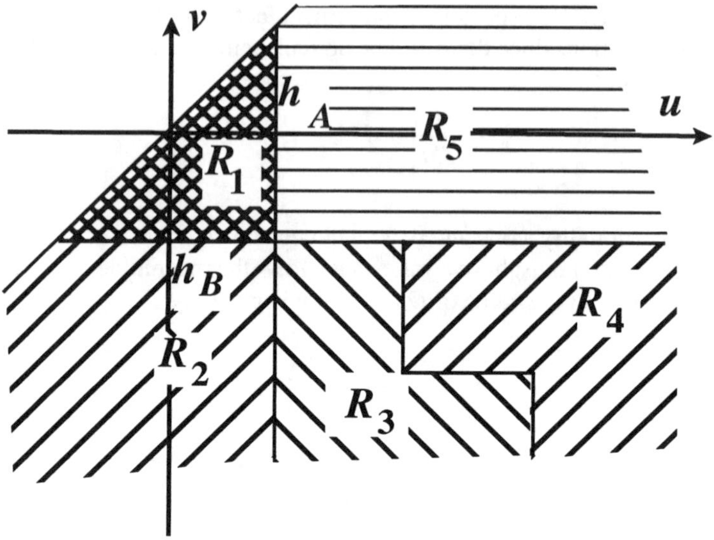

Figure 5.11 Description of minor loop behavior in the operative Preisach plane.

be equal to M_1. This height is independent of M. Furthermore, all the minor loops between the fields h_A and h_B are congruent in the operative plane, and the loops are stable immediately after a first application of h_A.

On the other hand, if this model is applied to the process illustrated in Fig. 5.11, when the field is h_A, then Q_1 and Q_2 will be +1. The value of Q_3, which would normally be +1, in this model will be less than 1, since it is diluted by hysterons coming in from R_4 and R_5. For similar reasons Q_4 and Q_5, that would normally be −1, will have a value somewhat greater than 1. The difference between the handling of Q_1 and Q_2 is that the former will oscillate between +1 and −1 and the latter will oscillate between 1 and a value only somewhat less than 1, when $h = h_B$. It is also noted that only Q_5 will have a value of +1 at that field.

We will consider only the part of a magnetizing process that comes after a suitable history has created the desired staircase on the Preisach plane. The first application of a field h_A will be called the *first iteration*, and this iteration number will be indexed after each successive leg of the magnetizing process. Thus, the first time that h equals h_A, each of the state variables $Q_1^{(1)}$ and $Q_2^{(1)}$ will be set equal to +1. The values of the other $Q^{(1)}$'s have a magnitude less than 1 determined by the magnetization history. It is noted that the value of $Q_3^{(1)}$ starts out positive and that the values of both $Q_4^{(1)}$ and $Q_5^{(1)}$ start out negative.

Since $|\Delta M_i^{(n)}|$ is equal to $p_1 S M_S$, we can use (5.41), to rewrite (5.45) as follows:

$$\Delta Q_j^{(n)} = \beta p_1 \left[\sum_{k=1}^{5} Q_k^{(n)} p_k - Q_j^{(n)} \right]. \tag{5.47}$$

We will use this equation to solve for the Q's at the conclusion of each leg of the magnetizing process. For even indices, the applied field is h_B, and we set

$$Q_1^{(2n)} = Q_5^{(2n)} = -1$$

$$Q_j^{(2n)} - Q_j^{(2n-1)} = \beta p_1(p_1 + p_2 + Q_3^{(2n-1)} p_3 + Q_4^{(2n-1)} p_4 - p_5 - Q_j^{(2n-1)}), \quad \text{for } j = 2, 3, \text{ and } 4. \tag{5.48}$$

We note that in this calculation we always reset $Q_2^{(2n-1)}$ equal to +1. For odd indexes, the applied field is h_A, and the Q's are given by

$$Q_1^{(2n+1)} = Q_2^{(2n+1)} = 1$$

$$Q_j^{(2n+1)} - Q_j^{(2n)} = \beta p_1(-p_1 + p_2 + Q_3^{(2n)} p_3 + Q_4^{(2n)} p_4 - p_5 - Q_j^{(2n)}), \quad \text{for } j = 3, 4, \text{ and } 5. \tag{5.49}$$

We note that in this calculation we always reset $Q_5^{(2n)}$ equal to -1. It is seen that if β is zero, then $Q_j^{(2n+1)} = Q_j^{(2n)} = Q_j^{(2n-1)}$, for $j = 3$ and 4, so there is no accommodation.

We see that the different regions have different roles in the accommodation process. Region R_1 is actively switched as the minor loops are traversed. Thus p_1 drives the accommodation process by forcing the hysterons to move in the Preisach plane. In alternate half-cycles, regions R_1 and R_5 suffer a small amount of accommodation, but then the magnitude of Q is restored to unity. The history of the magnetizing process is contained in R_3 and R_4. During the accommodation process, this history gradually fades away. If p_3 and p_4 are zero, as in the case of the major loop, then there is no history to be diluted and no accommodation of the end points of minor loops can take place, even if β is not zero. Finally, there are some minor loops for which no accommodation takes place. For example, no accommodation can take place when σ_k is zero, since if $(h_A - h_B)/2$ is greater than $\overline{h_k}$, then p_3 and p_4 are zero, but if $(h_A - h_B)/2$ is smaller than $\overline{h_k}$, then p_1 is zero.

The equilibrium minor loop, that is, the loop that finally closes on itself, can be computed by letting $Q_j^{(2n+2)} = Q_j^{(2n)} = Q^{(even)}$, for $j = 3$ and 4. Thus,

$$Q_j^{(\text{even})} = \frac{\dfrac{2(p_2 - p_5)}{p_1 + p_2 + p_5} - \beta p_1(-p_1 + p_2 - p_5)}{2 - \beta p_1(p_1 + p_2 + p_5)}, \tag{5.50}$$

and by letting $Q_j^{(2n+1)} = Q_j^{(2n-1)} = Q^{(odd)}$, for $j = 3$ and 4, we have

$$Q_j^{(\text{odd})} = \frac{\dfrac{2(p_2 - p_5)}{p_1 + p_2 + p_5} - \beta p_1(p_1 + p_2 - p_5)}{2 - \beta p_1(p_1 + p_2 + p_5)}. \tag{5.51}$$

The limiting magnetization is obtained from (5.41), so that at the upper end of the limiting minor loop we have

$$M_A = SM_S[p_1 + p_2 - Q_5^{(odd)} p_5 + Q^{(odd)}(p_3 + p_4)], \tag{5.52}$$

where

$$Q_5^{(odd)} = -1 + \beta p_1(1 - p_1 + p_2 - p_5 + (p_3 + p_4)Q^{(even)}), \tag{5.53}$$

while at the lower end of the limiting minor loop we have

$$M_B = SM_S[-p_1 + Q_2^{(even)} p_2 - p_5 + Q^{(even)}(p_3 + p_4)], \tag{5.54}$$

where

$$Q_2^{(even)} = 1 + \beta p_1(p_1 + p_2 - p_5 - 1 + (p_3 + p_4)Q^{(odd)}). \tag{5.55}$$

The height of the minor loop, $M_A - M_B$, will be slightly smaller than $2SM_S p_1$, the value it would have if there were no accommodation.

To illustrate the behavior of this model, let us consider the case $p_1 = 0.2$, $p_2 = 0.03$, $p_3 = 0.3$, $p_4 = 0.22$, $p_5 = 0.27$, and $\beta = 0.3$. For these specific values, Fig. 5.12 shows the variation in Q_3 and Q_4 as a function of the number of times a minor loop is traversed. It is seen that both curves exponentially approach the same limiting value asymptotically. Thus, accommodation is caused by the gradual disappearance of the staircase that divides the part of the Preisach plane that would

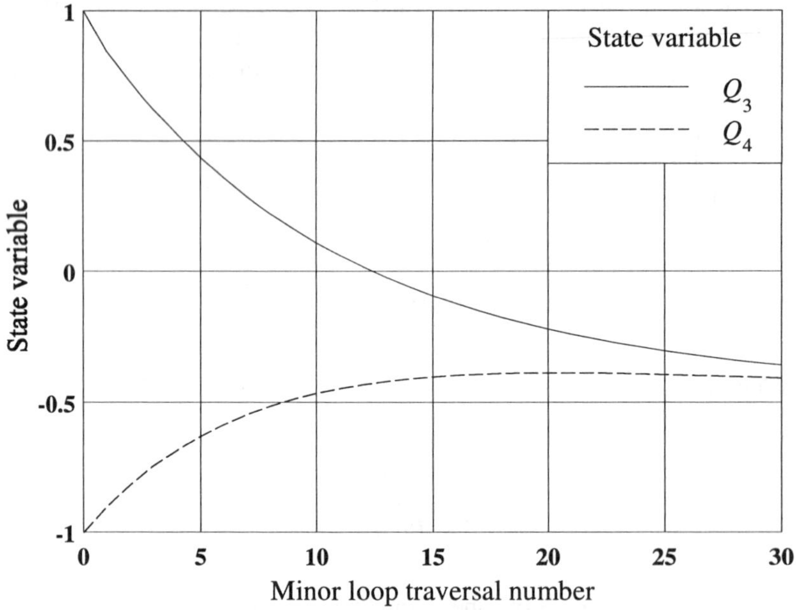

Figure 5.12 Change in state variables with number of minor loops traversed.

be unaffected by this process in other Preisach models. Furthermore, in this model, the actual structure of the staircase is immaterial. Only the values of the integrals of the Preisach function over each of the two areas are used.

The gradual shift in the minor loops can be seen by using (5.52) and (5.54) to calculate the magnetization at their ends as a function of the number of times a minor loop is traversed. For the same values of p and β, the variation in the end of the minor loops is shown in Fig. 5.13. Accommodation begins even at the first leg. Thus, in a nonaccommodating model, starting from negative saturation and applying a field of $\overline{h_k}$ would demagnetize the sample. In this model, for the same values, the magnetization will have a small positive value.

When one starts from negative saturation, the value for Q in the entire Preisach plane is equal to -1. When a field h is applied, the part of the Preisach plane to the left of h has a $Q = +1$. Thus, if hysterons from the right part of the plane move into the left part, they will experience a field sufficient to *correct* their magnetization. On the other hand, the part of the plane to the right of h will have a value greater than -1, since hysterons moving there will not have their magnetization corrected. That is, along this leg of the major loop, the value of Q in that part of the plane is a monotonically increasing function of h, which approaches a limit less than 1. When the applied field is large enough to saturate the material, the entire plane will achieve a value of +1 for Q. Thus, in agreement with experimental observations, there is no accommodation of the ends of the major loop as calculated by this

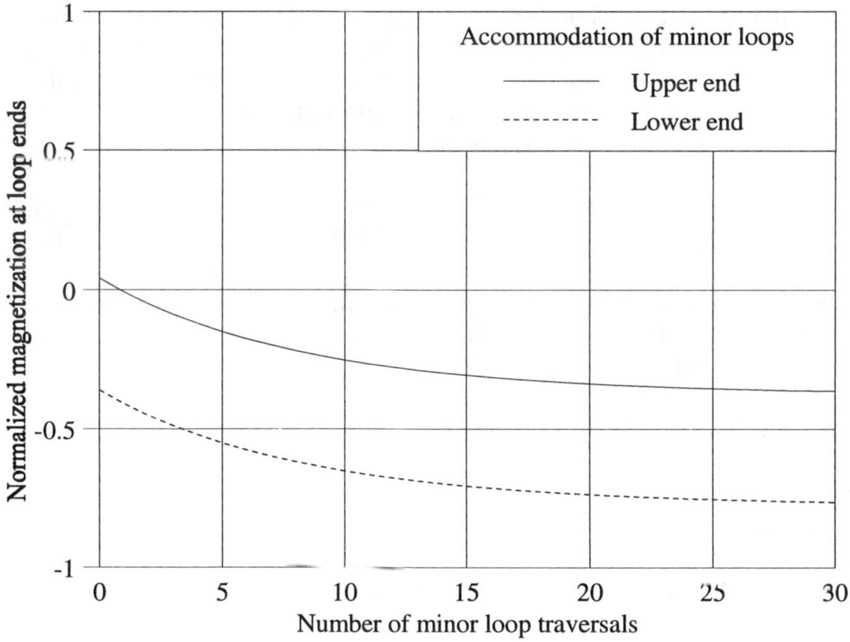

Figure 5.13 Magnetization accommodation of minor loops.

model.

When one is obtaining the end values of a minor loop for both large changes in the applied field and large values of β, there will be large errors if a leg of the process is traversed in a single step and a low order error method is used to solve the differential equations. Such a method is Euler's method, used above. More accurate methods, which have higher orders of error, such as the Runge–Kutta method and predictor-corrector methods, are discussed in standard numerical methods books.

An important problem in recording is the gradual decrease in the magnetization of a recording during successive playbacks. A major cause of this loss is the accommodation cycle caused by the playback head. A magnetized medium is subject to a demagnetizing field. This field is reduced when the medium is near or in contact with the playback head, since the medium acts as a keeper. Thus, an element of the medium repeatedly passed in contact with a head is subject to many minor loop cycles. These cycles range between effective fields that are the product of the element's magnetization and the two demagnetizing factors: one in the presence and one in the absence of the playback head. The most expedient way to reduce this decay in magnetization is to reduce p_1.

It is noted that in this analysis, unless p_1 is identically zero, accommodation will take place. This is an artifact of the approximation of a discrete particulate tape by a continuous Preisach function. In a real medium, the smallest entity that can be switched is the magnetization associated with a hysteron. Thus, if p_1 is less than that due to a single hysteron, it is for all practical purposes zero, and no accommodation occurs. Furthermore, if only a single hysteron is switched back and forth by this cycling, no accommodation will occur, since the original state of the interaction field is restored at the conclusion of the cycle. This latter extension can probably be extended to the switching of a few hysterons.

We note that in the limit as β approaches zero,

$$Q_j^{(\text{even})} = Q_j^{(\text{odd})} = \frac{p_2 - p_5}{p_1 + p_2 + p_5} . \tag{5.56}$$

These are the equilibrium values of Q that the minor loops try to achieve by accommodation; however, since there is no accommodation in this case, these values will never be achieved.

5.6 IDENTIFICATION OF ACCOMMODATION PARAMETERS

This model has only one new parameter, β, to be identified. The identification of the parameters of the CMH model has been possible from major loop data only [20], since these are not affected by accommodation. A way to identify this parameter is to measure the drift in a minor loop. To obtain the most accurate measure of β, it is necessary to obtain the greatest amount of accommodation. To maximize accommodation, one must simultaneously maximize p_1 (to maximize the motion of hysterons in the Preisach plane) and maximize either p_3 or p_4 (to maximize the magnitude of the magnetization that must be *forgotten*). For symmetrical Preisach functions, this is done by choosing h_A to be $\overline{h_k}$ and h_B to be $-\overline{h_k}$ for the extrema of the minor loop. For a nonaccommodating model, these fields would be the operative remanent coercivities; because of accommodation, however, the magnetization is not zero when the field is h_A. In that case, p_2 is equal to p_5 and p_1 is equal to $p_3 + p_4$. Furthermore, if we start from negative saturation, then p_3 is zero. In the subsequent calculations, we will assume that $h_A = \overline{h_k} = -h_B$.

The ratio of p_1 to p_2 is determined by the ratio of σ_k to σ_i; for example if $\sigma_i = 0$, then $p_2 = 0$, if $\sigma_i = \sigma_k$, then $p_1 = p_2$, and if $\sigma_k = 0$, then $p_1 = 0$. In the following analysis we will assume that this is the case. Then, since the p's are normalized, we have $p_1 = p_2 = p_4 = p_5 = 1/4$. Since the drift in the minor loops is small, it is possible to use the Euler method of solution discussed above to describe the accommodation. The value of β does not affect the equilibrium value of the minor loop, but it does affect the rate at which equilibrium is approached. At h_B, we use (5.48) to find that for even indices and when $j = 2$ and 4, the Q's are given by

$$Q_j^{(2n)} - Q_j^{(2n-1)} = \frac{\beta}{16}\left[1 + Q_4^{(2n-1)} - 4Q_j^{(2n-1)}\right], \tag{5.57}$$

while for odd indices, at h_A, when $j = 4$ and 5, the Q's are given by

$$Q_j^{(2n+1)} - Q_j^{(2n)} = \frac{\beta}{16}\left[-1 + Q_4^{(2n)} - 4Q_j^{(2n)}\right]. \tag{5.58}$$

For even n, (5.57) reduces to

$$Q_1^{(2n)} = Q_5^{(2n)} = -1, \quad Q_2^{(2n)} = \left(1 - \frac{3\beta}{16}\right) + \frac{\beta}{16}Q_4^{(2n-1)}, \tag{5.59}$$

and

$$Q_4^{(2n)} = \left(1 - \frac{3\beta}{16}\right)Q_4^{(2n-1)} + \frac{\beta}{16}, \tag{5.60}$$

and for odd n, (5.58) reduces to

$$Q_1^{(2n+1)} = Q_2^{(2n+1)} = 1, \quad Q_4^{(2n+1)} = \left(1 - \frac{3\beta}{16}\right)Q_4^{(2n)} - \frac{\beta}{16}, \tag{5.61}$$

and

$$Q_5^{(2n+1)} = \left(-1 + \frac{3\beta}{16}\right) + \frac{\beta}{16}Q_4^{(2n)}. \tag{5.62}$$

We note that

$$Q_4^{(1)} = Q_5^{(1)} = -1 + \frac{\beta}{8}. \tag{5.63}$$

Therefore, at the end of the first leg of the magnetizing process, the magnetization is given by

$$M_i = \frac{\beta S M_S}{16}. \tag{5.64}$$

We see that if β were zero, the magnetization would be zero, and we would be at the remanent operative coercive field.

The recursion relations for $Q_4^{(\text{even})}$ can be written

$$Q_4^{(2n+2)} = \left(1 - \frac{3\beta}{16}\right)^2 Q_4^{(2n)} + \frac{3\beta^2}{256}, \tag{5.65}$$

and for $Q_4^{(\text{odd})}$ can be written

$$Q_4^{(2n+1)} = \left(1 - \frac{3\beta}{16}\right)^2 Q_4^{(2n-1)} - \frac{3\beta^2}{256}. \tag{5.66}$$

Thus, two legs later in the magnetizing process, when the field is h_A again, $Q_4^{(3)}$ is given by

$$Q_4^{(3)} = -1 + \frac{\beta}{2} + \frac{159\beta^2}{256} + \frac{9\beta^3}{128}. \tag{5.67}$$

Since $Q_5^{(\text{even})}$ is always −1, then $Q_5^{(3)}$ is given by

$$Q_5^{(3)} = -1 + \frac{3\beta}{16} + \frac{\beta}{16} Q_5^{(2)} = -1 + \frac{\beta}{8}. \tag{5.68}$$

Then the magnetization will be

$$\frac{M_i}{SM_S} = \frac{9\beta}{8} + \frac{159\beta^2}{256} + \frac{3\beta^3}{128}. \tag{5.69}$$

For small values of β, we can neglect higher powers of β and approximate the magnetization by taking only the first term. This magnetization is larger than that of the first leg by approximately $\beta SM_S/8$, and thus the loop does not close. By comparing these two values of the magnetization at h_A we can obtain an estimate for β. Thus,

$$\beta \approx \frac{8\Delta M}{17 SM_S}. \tag{5.70}$$

If this value of β is too small to be measured accurately, a more appropriate formula can be derived using more cycles. For small β, we can again neglect higher order terms, and (5.66) can be written

$$\Delta Q_4^{(2n-1)} \approx -\frac{3\beta}{8} Q_4^{(2n-1)}. \tag{5.71}$$

The solution to this equation is

$$Q_4^{(2n+1)} \approx -\left(1 - \frac{\beta}{8}\right)\left(1 - \frac{3\beta}{8}\right)^{2n}. \tag{5.72}$$

It is seen that Q_4 goes from $(-1 + \beta/8)$ to zero, as n increases. The approach to equilibrium implied in this equation is the same exponential variation illustrated in Fig. 5.13.

It is noted that in these calculations, the operative fields of h_A and h_B were kept constant in the accommodation process and the applied fields were allowed to vary. This can be done on a vibrating sample magnetometer (VSM), especially on a programmable one, once the value of α is known. By measuring the magnetization as the field is applied, one can iteratively modify the applied field accordingly. Alternately, to keep the applied field limits constant during the accommodation process, it is necessary to derive new formulas, since the magnetization changes as the accommodation process develops.

A statistically derived Preisach model and some of its properties have been presented for the accommodation in minor loops. The model has been derived from a statistical interpretation of the physical principles underlying the Preisach model. In addition, a measurement technique has been suggested to calculate the parameter, β, introduced by this model. The identification process must be extended to the case where σ_i/σ_k is not unity, and the method of the identification of the accommodation parameter must be extended to include accommodation corrections.

Experiments have yet to be done to determine the applicability of this model. It is believed that this model is appropriate for longitudinal magnetic recording media that can be accurately described by the CMH model. For vertical media, a similar calculation based on the variable-variance model [21] must be derived. It is also suggested that a more sophisticated model might be necessary to fit experimental results. In the more sophisticated model, the state variable, Q, in a given region is not simply a constant, but a function of the critical field, h_k. This could be the case for a thin film medium that is perfectly aligned. Finally, it is hoped that this model, along with the aftereffect model, might be useful to determine the archivability of recordings.

5.7 PROPERTIES OF ACCOMMODATION MODELS

We can use these definitions and the notation and method of computing Q given in the preceding sections to compute the reversible and the irreversible components of the magnetization by generalizing the results obtained for the state-dependent reversible magnetization model [20]. A simplification results if we assume that the normalized reversible function can be factored into the product of a function of the applied field and a function of the interaction field. For example, if a branch of an isolated hysteresis loop can be written as

$$\begin{aligned} f(H-h_i)-f(h_i) &= f(H)g(h_i) \qquad \text{if } Q = 1 \\ f(-h_i)-f(-H+h_i) &= f(H)g(h_i) \qquad \text{if } Q = -1, \end{aligned} \tag{5.73}$$

then the irreversible component of the media magnetization is given by

$$M_i = SM_S\int_{-\infty}^{\infty}\int_{0}^{\infty} m_i(h_i)p(h_i,h_k)\,dh_k\,dh_i, \tag{5.74}$$

where

$$m_i(h_i) = Q[(1-S)f(-Qh_i)+S]. \tag{5.75}$$

The locally reversible, state-dependent component of magnetization is given by

$$M_r(H) = a_+ f(H) - a_- f(-H), \tag{5.76}$$

where the *reversible coefficient* is given by

$$a_\pm = (1-S)M_S\int_{-\infty}^{\infty}\int_{0}^{\infty}\frac{1\pm Q}{2}g(h_i)p(h_i,h_k)\,dh_k\,dh_i. \tag{5.77}$$

It is seen that for square loop materials, S is 1 and $m_i(h_i)$ reduces to Q. For nonaccommodating models, the magnitude of Q is unity and the term $(1 + Q)/2$ is one in regions that are magnetized positively and zero where they are magnetized negatively, thus, reducing to the definitions in the CMH model.

We will define the *regional reversible coefficients* by

$$a_{j\pm} = (1-S)M_S\iint_{R_j} g(h_i)p(h_i,h_k)\,dh_i\,dh_k, \tag{5.78}$$

where Q_j is the value of the state variable in region R_j. This definition depends only on the shape of the region. Thus, (5.76) still holds with the definition that

$$a_\pm = \sum_{j=1}^{5}\frac{1\pm Q_j}{2}a_{j\pm}, \tag{5.79}$$

which explicitly illustrates the state variable dependence of the locally reversible magnetization. Similarly, to illustrate the state dependence of the irreversible magnetization, we can define *regional irreversible coefficients* that depend only on the shape of the region by

$$p_j = \iint_{R_j}[(1-S)f(-Q_jh_i)+S]p(h_i,h_k)\,dh_k\,dh_i. \tag{5.80}$$

With this definition, we see that the sum of the p's is unity; therefore, using (5.74) we may rewrite (5.75) as follows:

$$M_i = SM_S \sum_{j=1}^{5} Q_j \, p_j. \tag{5.81}$$

In this analysis, we will examine only the end points of the minor loops and study their drift. Hence, for a process that oscillates between the same two operative fields, the five regions in Fig. 5.11 are stationary and the integrals in (5.79) and (5.80) are constant at the limits of the magnetization cycles. Thus, the only drift will be due to the changing values of the Q.

5.7.1 Types of Accommodation Processes

An accommodation process occurs when a magnetic medium is cycled between two values of applied field. We will define three types of accommodation process: *operative field accommodation* (OFA), *applied field accommodation* (AFA), and *demagnetizing factor accommodation* (DFA).

In OFA, the magnetization is cycled between a pair of operative fields; that is, the applied fields are changed by $\alpha\Delta M$ whenever the magnetization changes by the amount ΔM. To be able to apply an operative field, one must measure the magnetization as the process proceeds, and iteratively and monotonically correct the applied field until the desired operative field is attained. It is suggested that this process may be used to identify the accommodation parameter, since the Preisach functions are stable in the operative plane. When discussing cycling between fields h_A and h_B, we will refer to the point (h_A, h_B) in the operative plane as the *operating point.*

In AFA, the magnetization is cycled between a pair of applied fields. This is the easiest type of accommodation process to perform experimentally, since the applied field just oscillates between a pair of field extrema. It is more difficult to interpret than the OFA process because as the magnetization accommodates, the operating point moves. Furthermore, the minor hysteresis loops change during the accommodation process because of the lack of the congruency property in the medium.

The final accommodation process, DFA, occurs whenever the geometry of the magnetic circuit changes (e.g., upon the application and subsequent removal of a keeper). This process also occurs whenever a recording head passes over a recording medium. In these cases the demagnetization factor changes with the geometrical changes, thus effectively cycling the demagnetizing field. Just as in the AFA process, the operating point moves during accommodation, but, since the applied field depends on the magnetization, it changes as well.

We will first consider an operative field accommodation process, since the limits of the minor loop excursions are constant in the operative plane and thus simpler to describe. For a given medium, the amount of accommodation depends on the values of h_a and h_b, as well as the magnetization history. The loop will drift with each subsequent cycle and eventually reach a stable loop, which we will call the *equilibrium loop*. Since accommodation wipes out the magnetization history, the equilibrium loop is only a function of h_A and h_B; however, the way this loop is approached does depend on the magnetization history. It has been observed that major loops do not accommodate. Since there exist media that do not spontaneously demagnetize, there must be a threshold field below which no accommodation occurs. We will now discuss these and other properties of the accommodation model.

The limit fields h_A and h_B define a point on the Preisach plane that divides this plane into four regions: R_1, R_2, R_{34}, and R_5. The region R_{34} is the combination of R_3 and R_4. With this division, the regions R_1 and R_2 are magnetized positively when the applied field is h_A and the regions R_1 and R_5 are magnetized negatively when the applied field is h_B. The region R_{34} is unaffected directly by this process; however, the motion of the hysterons in the Preisach plane, causes the magnetic state of this region to tend to become homogeneous.

For this accommodation process, the most positive value the magnetization can take is found when R_{34} is initially magnetized positively and the applied field is h_A. At this point the minor hysteresis loop is near the upper branch of the major loop and Q_{34} is almost 1. During subsequent cycles, Q_{34} will decrease and the loop will drift downward. Similarly, the most negative value the magnetization can take occurs when initially R_{34} is magnetized negatively and the applied field is h_B. At this point the minor hysteresis loop is near the lower branch of the major loop, and Q_{34} is almost -1. Thus, in this model, a minor loop will always lie inside the major loop. Furthermore, the maximum accommodation that can take place is at the point where the major loop is widest, and that occurs for loops where $h_A = -h_B$. These loops will be referred to as *symmetrical minor loops*.

The size of the first drift in the positive end of a minor loop is proportional to the product of p_1 and p_{34}. For symmetrical minor loops, p_1 is a monotonic increasing function of h_A starting from zero when h_A is zero, and p_{34} is a monotonic decreasing function of h_A that goes to zero for large h_A. Therefore, their product starts at zero and will go through a maximum as h_A is increased from zero. It can be shown that the maximum occurs at the operative remanence coercivity. On the other hand, for major loops, R_{34} is zero, and there is no accommodation of the end points of the loop.

For symmetrical minor loops, the equilibrium loop will also be symmetrical in the magnetization as well as the operative field. Since the magnetization at the two ends of the minor loop are equal in magnitude but opposite in sign, the minor loop will be symmetrical with respect to the applied field as well. That is, the center of the equilibrium loop will be the origin. Thus, in an *ac* demagnetization process, it is not necessary to have a field large enough to saturate the sample to delete the magnetization history, but simply to go through a sufficient number of cycles before the applied *ac* field is reduced to zero. It can be shown that for this model, these two demagnetization processes and the Curie point demagnetization produce the same magnetization sequence for the same applied field sequence.

It should be pointed out that this model has one other property: all accommodating minor remanence loops lie within the major remanence loop, and their equilibrium position lies at the midpoint of the section of the major loop between the two field limits. This can be seen from the fact that m_i on the ascending major remanence loop at h_A is given by

$$m_{i\,\mathrm{asc}}(h_A) = SM_S(p_1 + p_2 - p_{34} - p_5). \tag{5.82}$$

For any minor loop, since the magnitudes of all the q's are less than one, it is seen that $m_i\ (h_A)$ is given by

$$m_i(h_A) = SM_S(p_1 + p_2 - Q_{34}p_{34} - p_5) > m_{i\ \mathrm{asc}}(h_A). \tag{5.83}$$

Thus, the right ends of all minor loops lie above the ascending major remanence loop. Furthermore, the descending major remanence loop magnetization at h_A is given by

$$m_{i\,\mathrm{desc}} = SM_S(p_1 + p_2 + p_{34} + \nu p_5), \tag{5.84}$$

where ν, a positive fraction that is less than 1, is the fraction of R_5 that is still positive when the applied field is reduced to H_A. Furthermore, for Preisach functions that are limited to the fourth quadrant, if H_A is positive, then ν is 1. Comparing with (5.83), it is seen that this is greater than $m_i(h_A)$. Therefore, the right end of the minor loop also lies below the descending major remanence loop. Since the reversible component in the CMH loop is also largest for the major loop, the analysis above can be extended to the total magnetization. By similar reasoning, it can be shown that the left end of minor loops lie above the ascending major remanence loop.

For small β the equilibrium loop, the state variable Q_{34} is given by

$$Q_{34} = \frac{p_2 - p_5}{p_1 + p_2 + p_5}. \tag{5.85}$$

Thus, it can be shown that the average magnetization for the equilibrium loop is given by

$$\langle M_i \rangle = SM_S \frac{p_2 - p_5}{p_1 + p_2 + p_5}. \tag{5.86}$$

This magnetization is the average of the magnetization of the region that is affected by the applied fields and generally lies in the center of the major remanence loop. Therefore, minor loops starting at the major loop will accommodate away from the major loop. Furthermore, this limiting average magnetization is zero for symmetrical minor loops. When there is cycling between two applied fields, the operating point changes with the magnetization, as illustrated in Fig. 5.14. If the process observes the congruency property, the locus of operating points is a straight line with unit slope.

The DFA process occurs whenever the geometry changes (e.g., when a keeper is brought up to a permanent magnet or when a recorded medium is passed near a recording head). If this activity is repeated cyclically, accommodation can take place, and in certain cases the medium can become demagnetized. In this case, the field in a magnetized medium changes because the geometric demagnetization factors change. Thus, if the geometry is cycled, the medium will experience a cyclical applied field. This field is similar to the AFA process except that both field limits are now of the same sign and as the magnetization accommodates, the field limits change.

It is seen that in this case, the "applied field" is only of one sign. Thus, at equilibrium the average magnetization will not be zero; however, it can become very small. It is important that this limit be calculated, since if it is too small the medium will be useless for recording. This phenomenon limits the usefulness of media for all wavelengths of recording, but for short wavelength recording there is another effect.

In longitudinal digital magnetic recording, ideally the medium is magnetized to saturation in alternating directions in regions separated by an abrupt transition that is perpendicular to the track. In a real medium this transition occurs over a finite distance, and a detailed plot of the magnetization along the track is shown in Fig. 5.15. However, DFA may cause the medium to be slightly demagnetized as shown, thus resulting in a more gradual transition. This is one of the limits associated with recording density.

This accommodation process [5] is able to describe accommodation with only a single new parameter. It is able to predict what a stable minor loop would be as a function of the limits of the applied field excursions. Preliminary measurements [3] have shown that it appears to describe the general features of accommodation.

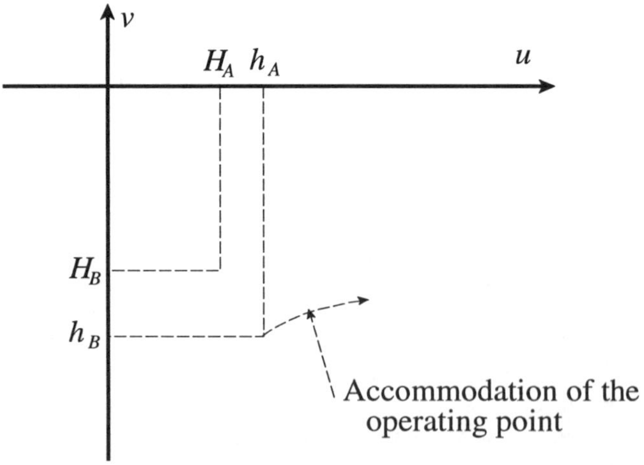

Figure 5.14 Motion of the operating point during an AFA process.

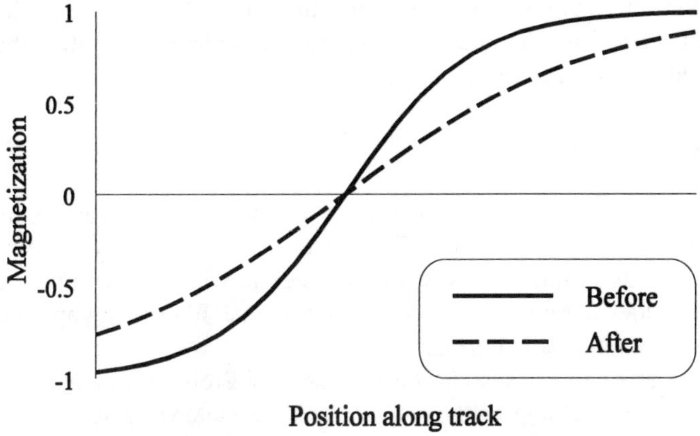

Figure 5.15 Transition broadening in longitudinal digital magnetic recording due to accommodation.

The features of this model are as follows: The major loop does not accommodate. Minor loops always lie inside the major loop. Minor loops accommodate away from the major loop. The magnetization is stable if the applied field does not change. Accommodation distorts the symmetry of all loops, and if hysteron interaction decreases, accommodation decreases.

5.8 DELETION PROPERTY

In Chapter 2 we saw that the deletion property of the Preisach model was directly related to the uniqueness of the Everett integral as a description of the magnetization change. The proof of the deletion property was based on the assumption that changes in magnetization are completely determined by this Everett integral. This is no longer the case when there is accommodation, aftereffect, or both.

Whenever a field is applied, the Preisach plane is divided into three regions: the two regions where the field determines the magnetic state of the hysterons, and a region where the hysteron can be in either state. It is this latter region—also called the *unaffected region*, since it would not be affected by the magnetizing process in the classical Preisach model—that causes the violation of the deletion property. For the state to be determined by the Everett integral, it is necessary for the state vector to be constant in this region; however, it can be shown from (5.31) and (5.46) that the state vector in this region obeys the following differential equation:

$$\frac{dQ(u,v)}{dt} = \frac{\tau_+ - \tau_-}{\tau_+\tau_-} - Q(u,v)\frac{\tau_+ + \tau_-}{\tau_+\tau_-} + \frac{\beta[M_S - SM_SQ(u,v)]}{S^2M_S^2}\frac{d|M|}{dt}. \tag{5.87}$$

The time derivative of the magnetization is the sum of the integral of this function over the unaffected region, plus the magnetization changes for the other regions, as computed in the preceding chapters.

5.9 CONCLUSIONS

The gradual drift of minor loops can be rate independent due to accommodation or rate dependent due to aftereffect. The first will vary with cycle number when executing repeated minor loops, while the second will drift with time even if the applied field does not change. If one applies a small alternating applied field, the two methods can be easily confused.

Both types of processes can be modeled by Preisach models and relax the deletion property by changing the magnitude of the state variable. A new parameter β must be introduced to model accommodation, and two new parameters τ h_f must be introduced to model accommodation.

REFERENCES

[1] S. Chikazumi with S. H. Charap, *Physics of Magnetism*, Wiley: New York, 1964.

[2] J. L. Snoek, "Time effects in magnetization," *Physica*, **5**, 1938, pp. 663–688.

[3] Y. Tomono, "Magnetic after effect of cold rolled iron, I," *J. Phys. Soc. Japan*, **7**, 1952, pp. 174–179.

[4] F. Preisach, "Über die magnetische Nachwirkung," *Z. Phys.*, **94**, 1935, pp. 277–302.

[5] A. Aharoni, *Introduction to the Theory of Ferromagnetism*, Clarendon Press: Oxford, 1996.

[6] C. Korman and I. D. Mayergoyz, "Preisach model driven by stochastic inputs as a model for aftereffect," *IEEE Trans. Magn.*, **MAG-32**, September 1996, pp. 4204–4209.

[7] G. Bertotti, "Energetic and thermodynamic aspects of hysteresis," *Phys. Rev. Lett.*, **76**, 1996, pp.1739–1742.

[8] E. Della Torre and L. H. Bennett, "A Preisach model for aftereffect," *IEEE Trans. Magn.*, **MAG-34**, July 1998, pp. 1276–1278.

[9] Y. D. Yan and E. Della Torre, "Particle interaction in numerical micromagnetic modeling," *J. Appl. Phys.*, **67(9)**, May 1990, pp. 5370–5372.

[10] G. Bottoni, "Size effect on the time dependence of magnetization of iron oxide particles," *IEEE Trans. Magn.*, **MAG-33**, September 1997, pp. 3049–3051.

[11] G. Bottoni, D. Candolfo, and A. Cecchetti, "Interaction effects of the time dependence of the magnetization in recording particles," *J. Appl. Phys.*, **81**, 1997, pp. 3809–3811.

[12] E. P. Wohlfarth, "The coefficient of magnetic viscosity," *J. Phys. F: Met. Phys.*, **14**, August 1984, L 155–L159.

[13] R. Street and J. C. Woolley, "A study of magnetic viscosity," *Proc. Phys. Soc.*, **A 62**, 1949, pp. 562–572.

[14] M. Lo Bue, V. Basso, G. Bertotti, and K.-H. Müller, "Magnetic aftereffect in spring magnets and the Preisach model of hysteresis," *IEEE Trans. Magn.*, **MAG-33**, September 1997, pp. 3862–3864.

[15] (a) E. Della Torre, L. H. Bennett, and L. J. Swartzendruber, "Modeling complex aftereffect behavior in recording materials using a Preisach-Arrhenius approach," *Mat. Res. Soc. Symp. Proc.*, **517**, 1998, pp. 291-296. (b) L. J. Swartzendruber, L. H. Bennett, E. Della Torre, H. J. Brown, and J. H. Judy, "Behavior of magnetic aftereffect along a magnetization reversal curve in a metal particle recording material," *Mat. Res. Soc. Symp. Proc.*, **517**, 1998, pp. 360-366.

[16] M. Brokate and E. Della Torre, "The wiping-out property of the moving model," *IEEE Trans. Magn.*, **MAG-27**, September 1991, pp. 3811–3814.

[17] E. Della Torre and G. Kádár, "Hysteresis Modeling II: Accommodation," *IEEE Trans. Magn.*, **MAG-23**, September 1987, pp. 2823–2825.

[18] O. Benda, "Possibilities and limits of the Preisach model," *J. Magn. & Magn. Mater.*, **112**, 1992, pp. 443–446.

[19] I. D. Mayergoyz, *Mathematical Models of Hysteresis*, New York: Springer-Verlag, 1991, p. 108.

[20] E. Della Torre and F. Vajda, "Parameter identification of the complete-moving hysteresis model using major loop data," *IEEE Trans Magn.*, **MAG-30**, November 1994, pp. 4987–5000.

[21] F. Vajda, E. Della Torre, M. Pardavi-Horvath, and G. Vertesy, "A variable variance Preisach model," *IEEE Trans. Magn.*, **MAG-29**, November 1993, pp. 3793–3795.

CHAPTER 6

VECTOR MODELS

6.1 INTRODUCTION

So far we have been discussing increasingly accurate scalar models for the magnetizing process. We can think of these as processes in which all the field variations lie along an axis, and we are interested only in the component of the magnetization along that axis. In a real magnetizing process, besides changing its value, the applied field could rotate. Furthermore, especially if the material is not isotropic, the resulting magnetization might not be in the same direction as the applied field. Thus, it is necessary to characterize material behavior in two or more dimensions. In this chapter we discuss how the work of Chapters 1 to 5 can be extended into two- and three- dimensional situations.

Before we address specific models, we will identify the general properties of vector models that are physically realizable. Besides the limits imposed on the scalar models, we will add two more properties. The *saturation property* refers to the requirement that all magnetizations calculated by the model not exceed saturation. The *loss property* refers to the fact that as the size of a rotating field increases, the losses first increase and then decrease. Both properties can be achieved by vector models.

We discuss three types of vector models. The *Mayergoyz vector model* is a purely phenomenological extension of the scalar Preisach models. On the other hand we can construct *pseudoparticle models* based on micromagnetic models, such as the Stoner–Wohlfarth model. These models can require substantial computation intensity. A middle course is the *coupled-hysteron* model, which couples three scalar models to obtain three-dimensional vectors, and adjusts them so that they satisfy the general requirements for vector models.

6.2 GENERAL PROPERTIES OF VECTOR MODELS

When the magnetization changes in a magnetic material, energy may be dissipated by various causes. It is convenient to categorize these energy losses as static losses and dynamic losses. The static losses are those that would occur when the magnetization is cycled arbitrarily slowly; the dynamic losses, some of which are discussed in Chapter 7, are the additional losses that occur when the magnetization is cycled more quickly and are a function of how quickly the magnetization varies with time. Static losses are caused by sudden changes in magnetization, when a field threshold is exceeded, such as those due to Barkhausen jumps.

In earlier chapters we discussed how these effects can be modeled to varying degrees of accuracy by various scalar models, for applied fields acting along a single axis. When the applied field changes its direction as well as its magnitude, the modeling becomes more complicated. Several vector extensions of Preisach models have been proposed in the last decade.

One of the properties a vector model should have is the *saturation property*: that is, for a large applied field in any direction, the magnetization should never exceed saturation. Furthermore, it should be able to achieve saturation, and for any direction of the applied field, by means of the application of a sufficiently large field. Then, as long as the field is applied, the magnetization should be in the same direction as the field. Thus, for a large rotating field, the locus of magnetization vector tips should trace out a circle. The three types of models discussed in the following sections all have this property.

We now describe some of the energy loss properties vector models should possess, and discuss how these models may be modified to achieve the desired loss variation with the applied field. We will concentrate on two such vector loss mechanisms in magnetic materials: that associated with anisotropy and that associated with wall motion. Other types of rotational loss mechanisms have been observed, but these are beyond the scope of this work. Since the models behave differently under a rotating field whose magnitude is increasing, this property can be used to distinguish between the various proposed models.

When an increasing oscillating field is applied to a magnetic material, the energy loss per cycle due to hysteresis is zero until a threshold field is reached. Then the loss increases until the material is saturated. Any further increase in the magnitude of the field does not increase or decrease the static loss per cycle. For both types of rotational hysteresis loss, the situation is different when the material is subject to an increasing rotating magnetic field.

The first type of rotational hysteresis, called *anisotropy* hysteresis, occurs in single domain particles when the magnetization attempts to follow a rotating applied field but is prevented from doing so by either shape or magnetocrystalline anisotropy. This type of hysteresis is characterized by a zero loss for small fields, which first increases and then decreases to zero as the applied field is increased. The analytic properties of this loss can be derived by considering the Stoner–Wohlfarth model for uniformly magnetized ellipsoidal particles.

The second type of rotational hysteresis, called *wall motion* hysteresis, occurs in materials that are large enough to support multidomains. When two adjacent domains, separated by a domain wall, have different orientations, then the domain whose orientation is closer to the applied field will grow at the expense of the other. As the applied field rotates, the direction of wall motion can even change. In these cases, the loss mechanism is due to Barkhausen jumps in wall motion, when domains with lower Zeeman energy grow at the expense of those with higher energy. Then the hysteresis loss for fields smaller than the minimum required to produce a Barkhausen jump is zero. As the field increases above this threshold, the loss increases as larger regions of the material are traversed by the domain walls. For fields large enough to saturate the material, the loss again decreases to zero, since all domain walls are eliminated. The range of fields for which hysteresis loss is present is much larger for these effects than for anisotropy hysteresis.

Thus, in both these cases, as the rotating magnetic field increases, the energy loss per cycle due to hysteresis is essentially zero until a threshold field is reached. Then the loss increases until the contribution of the new thresholds is less than the decreasing effect due to the thresholds that have been previously exceeded. At that point, unlike the case of an oscillating field, the loss starts to decrease to zero as the material saturates.

In particular, a model for anisotropy hysteresis is the uniform magnetization model for an isolated spheroidal magnetic particle, proposed by Stoner and Wohlfarth and discussed in Chapter 1. When the energy loss is plotted as a function of the applied rotating field, one obtains a curve as shown in Fig. 6.1. It is seen that there is no energy loss for applied fields that are less than the threshold required to change the state of the particle, since the process is entirely reversible. When the threshold is exceeded, the loss suddenly increases and then monotonically decreases with the applied field until it is reduced to zero. Further increases in the applied field, as is well known, do not produce losses, since for large fields, the magnetization is able to follow the applied field.

For an array of particles, although each particle behaves essentially in this way, the threshold field will be different for each particle. Furthermore, particle interaction may result in different magnitudes for the positive and negative switching fields. Nevertheless, as the rotating field is increased in magnitude, the loss will at first increase monotonically. At a critical field, the increase in loss associated with the switching of additional particles is equal to the decrease in loss of the particles with smaller critical fields. At this point the loss will decrease with

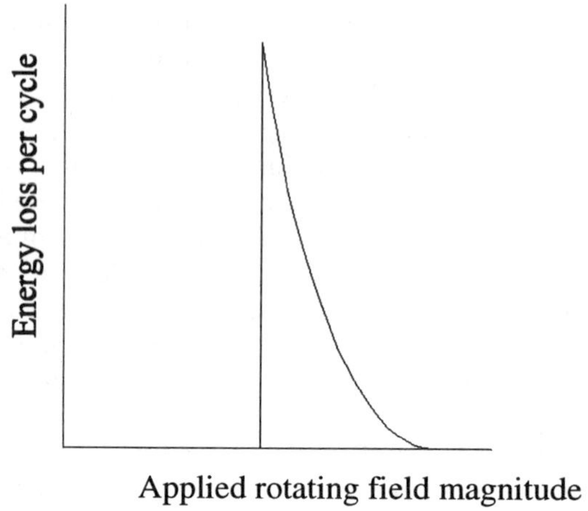

Figure 6.1 Rotational energy loss per cycle for a Stoner–Wohlfarth particle.

increasing field magnitude until all the particles are following the applied field. This is in sharp contrast with the loss associated with an alternating field that increases monotonically to saturation with the applied field.

The variation of the threshold field with the angle that it makes with the easy axis is fairly complicated for anisotropy hysteresis. In particular for a Stoner–Wohlfarth particle, the switching field variation with the angle of the applied field is an asteroid, discussed in Chapter 1. For a real particle, the angular variation is much more complicated. For wall motion hysteresis, on the other hand, the energy that the applied field supplies to the domain wall, to overcome the energy threshold, is the Zeeman energy. This energy varies as the cosine of θ, and the threshold field varies as its reciprocal; that is, as the secant of θ. Thus to make a reasonable model for the vector interpretation of the threshold field it is necessary to know the orientation of the easy axis and the mechanism of hysteresis. Since domain patterns in unsaturated specimens are random, even if their magnetization history is known, such an analysis must be statistical.

Thus, a vector model for hysteresis must be able to describe these effects. That is, it must reduce to the scalar model under the appropriate conditions, and in addition must obey the saturation property and the loss property in order to be physically realizable. Once it is physically realizable, the model should reproduce observed measurements. One of these results is the *remanence loop*, which is the locus of points traced out by the vector remanence as the direction of the applied field causing it is rotated. This remanence loop for many materials is an ellipse, and these materials are called *ellipsoidally magnetizable*. The major axis of that ellipse is the easy axis, and the minor axis is the hard axis. For isotropic media, the remanence loop is a circle.

6.3 THE MAYERGOYZ VECTOR MODEL

Mayergoyz proposed building a vector model from a continuum of scalar Preisach transducers [1], each incrementally rotated from its neighbor. The input to each transducer is the component of the applied field in that direction, and the output of each transducer is a magnetization in that direction. The output of the complete model is the vector sum of the output of all the transducers. Since his basic building block is a Preisach transducer, he quickly shows that his model reduces to the scalar Preisach transducer for processes that have a unique line of action. Furthermore, his model has the generalized congruency property; that is, for all cyclic magnetizing processes, the magnetization is also cyclic and the loops thus formed are all congruent to each other.

The Mayergoyz vector model computes the irreversible component of the magnetization as

$$\mathbf{m_i} = \iint_{u>v}\int_0^{\pi} p(\theta, u_\theta, v_\theta)\,\mathbf{Q}(\theta, u_\theta, v_\theta)\,d\theta\,du_\theta\,dv_\theta, \tag{6.1}$$

where $\mathbf{Q}$ is a unit vector lying either along the $\mathbf{1}_\theta$ or the $\mathbf{-1}_\theta$ direction. For isotropic media, the Preisach function $p(\theta, u_\theta, v_\theta)$, does not vary with θ. If a large field is applied along the line $\theta = \pi/2$, then $\mathbf{Q}$ lies along $\mathbf{1}_\theta$ for all θ. Then it is seen that $\mathbf{m_i}$ is in the $\mathbf{1}_\theta$ direction, and the components perpendicular to that direction cancel out in pairs. For anisotropic media, $p(\theta,u_\theta,v_\theta)$ varies with θ. Then it is seen that if we apply a large field and rotate it, the magnitude of the magnetization will vary. Moreover, its direction will not normally be in the same direction as the applied field, but will always make an acute angle with respect to it.

For smaller fields, the irreversible component of the magnetization will depend on the magnetizing history, since the medium is hysteretic. Each hysteron can have a different history, because it experiences a different sequence of applied fields. Thus, a different "staircase" must be stored for each hysteron.

The identification of isotropic media is comparatively simple, since all the hysterons are identical. Then all one has to do is to identify a typical hysteron; when a field is applied to one hysteron, however, the other hysterons may experience different fields. So even if they are identical, they will have different magnetization histories. We illustrate the identification process for two-dimensional processes. For simplicity, if one applies a field H in the direction $\theta = 0$, which will be taken as the x-axis, then hysterons in the direction θ will experience a field $H \cos \theta$. If one considers a first-order reversal process starting from a large negative value, going to a field H_1 and then to a smaller field H_2, the resulting magnetization is given by

$$\mathbf{m_i} = \mathbf{1_x}\int_0^{\pi} d\theta \int_{-\infty}^{H_1} dv_\theta \int_{H_2}^{H_1} du_\theta \cos\theta\, p(u_\theta \cos\theta, v_\theta \cos\theta)\,. \tag{6.2}$$

Differentiating with respect to u_θ and v_θ gives

$$\frac{\partial^2 \mathbf{m_i}}{\partial u_\theta \partial v_\theta} = \mathbf{1_x} \int_0^\pi \cos\theta \, p(u_\theta \cos\theta, v_\theta \cos\theta). \tag{6.3}$$

Unlike the case of the scalar model, the second partial derivative of the magnetization at the conclusion of a first-order reversal process does not yield the Preisach function directly.

Mayergoyz suggests two methods [1] to obtain the Preisach function from (6.3). The first method involves the evaluation of polynomial coefficients if (6.3) can be approximated by a polynomial. The second method involves a simple transformation that converts the integral equation into one of the Abel type. For anisotropic media, one must measure the magnetization for first-order reversal processes at all angles. The Preisach function is then obtained in terms of spherical harmonics. It is easy to show that this model has the saturation property, since the magnetization that it computes is always bounded. Therefore, if the saturation magnetization is set to be the least upper bound of these values, one can never exceed saturation.

6.4 PSEUDOPARTICLE MODELS

The pseudoparticle models approximate a hysteron by a small number of basic particles that are combined into a so-called *pseudoparticle*. Two such models have been proposed by Oti: one uses the Stoner–Wohlfarth model for the basic particles [2]; the other uses the results of a micromagnetic calculation for the basic particles [3]. Although the models assume that the hysterons are particles, their result can easily be extended to granular media. To illustrate how they work, let us assume that the pseudoparticle consists of three identical basic particles, as shown in Fig.

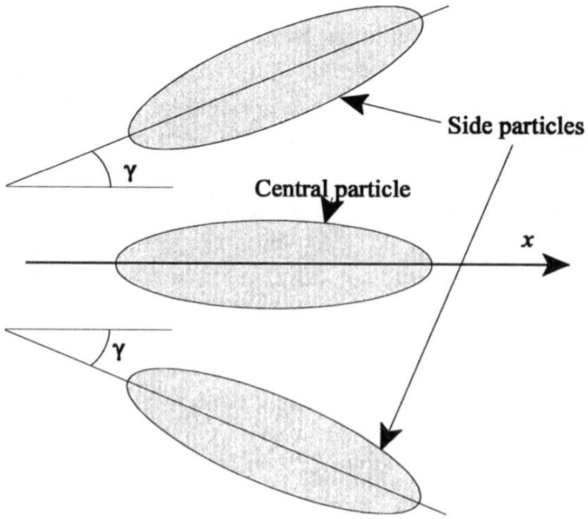

Figure 6.2 A pseudoparticle consisting of three basic particles.

6.2. If higher accuracy is desired, one can easily extend this model to include more basic particles.

We assume that the x axis, also called the *PMA* (Preisach measurement axis), is the easy axis of the medium and that the size of the moment and the angle made by the two side particles with the easy axis are the same. We therefore, have three independent variables: the moment of the central particle, the moment of one of the side particles, and the angle of the side particles. We can solve for these variables by requiring the pseudoparticles to have the same squareness as the medium as a whole, along three directions: the x direction, the y direction, and at an angle, say 45°, with respect to these axes.

If we call the moment of the central particle m_1, and of each of the side particles m_2, and the angle that each of the side particles makes with respect to the x axis θ, then the x squareness S_x is given by

$$S_x = \frac{m_1 + 2m_2\cos\theta}{m_1 + 2m_2}. \tag{6.4}$$

Similarly, the y squareness S_y is given by

$$m_2 = \frac{2m_2\sin\theta}{m_1 + 2m_2}. \tag{6.5}$$

If we apply a large field at other angles, we will find that the remanence is not in the same direction as the applied field. In particular, if θ is 45° and the applied field is also at 45°, we can assume that the lower of the two side particles is on the average demagnetized. The vector remanence of the pseudoparticle at zero field then is

$$\mathbf{m} = \left(m_1 + \frac{m_2}{\sqrt{2}}\right)\mathbf{1_x} + \frac{m_2}{\sqrt{2}}\mathbf{1_y}. \tag{6.6}$$

Thus, by changing θ, we can change the magnetization properties at other angles and thereby the shape of the remanence loop.

Each of the basic particles contains the angular variation of the process; however, each particle also represents a distribution of critical fields. Thus, the state of a particular basic particle is computed by a Preisach process. The Preisach distribution can be a normal distribution, and a moving model can be used to account for the variation in the local field with magnetization. Aftereffect and accommodation can also be introduced into this model, as discussed in Chapter 5. The identification of the Preisach parameters for each basic particle can be performed as a generalization of the identification of the scalar Preisach model. If we assume that the basic particles are identical, then once θ is known, we can project the effect of the two side particles on the PMA and use scalar identification on the composite particle.

Since the basic particles behave like real particles, in the case of the micromagnetic model or for small particles using the Stoner–Wohlfarth model, the system will naturally have the correct rotational properties. In particular, the system will have the saturation property and the loss property. The net magnetization of the system is obtained by taking the vector sum of the magnetization of the basic hysterons. It is noted that the Stoner–Wohlfarth model naturally computes the total magnetization of the hysteron. Hence, it is unnecessary to decompose the magnetization into a reversible and an irreversible component. Although in this model, we must maintain the magnetization history of only a few hysterons, in comparison to the many hysterons in the Mayergoyz model, since each basic particle produces a correct spatial field variation, the premise of the pseudoparticle model may be no less accurate.

6.5 COUPLED-HYSTERON MODELS

Another category of vector models consists of the coupled-hysteron models [4]. In this case we place a Preisach model along the principal axes of the system: two for two-dimensional models and three for three-dimensional models. If these models are permitted to be independent, the saturation property can easily be violated. The coupling is accomplished through a combined Preisach function. For example, for three-dimensional models, there are six Preisach variables: the up- and down-switching fields in the x, y, and z directions, respectively. The magnetization is computed by means of

$$\mathbf{m}_{\mathrm{i}} = \int_{\Omega_R} \cdots \int \mathbf{Q}(u_x, v_x, u_y, v_y, u_z, v_z) p(u_x, v_x, u_y, v_y, u_z, v_z)\, du_z\, dv_z\, du_y\, dv_y\, du_x\, dv_x. \tag{6.7}$$

where Ω_R is the region where $u_x > v_x$, $u_y > v_y$ and $u_z > v_z$. By requiring that **Q's** be less than or equal to one, we guarantee that the magnitude of $\mathbf{m}_{\mathrm{i}}$ is always less than one. We will define the composite Preisach volume as the six-dimensional hypervolume whose axes are u_x, v_x, u_y, v_y, u_z, and v_z . A point in this six-dimensional space will be denoted simply by Ω, so that this equation can be written

$$\mathbf{m}_{\mathrm{i}} = \int_{\Omega_R} \mathbf{Q}(\Omega) p(\Omega)\, d\Omega. \tag{6.8}$$

6.5.1 Selection Rules

The selection of **Q**, the state vector, is determined by selection rules. For simplicity, in this section, we assume that the x axis is the easy axis and the y and z axes are relatively harder axes of the material. Then the applied field will be decomposed into the x-direction, the y-direction components, and the z components. Models will then be built to compute the corresponding components of the magnetization. This will avoid cross terms in the calculations.

The state vector at a point in the Preisach volume represents the average state of a group of hysterons that have the same switching field but may have different orientations, size, shape, etc. Two such hysterons are indicated schematically in Fig. 6.3. When a large horizontal field is applied, the horizontal component of their magnetization will become positive. In that case, their vertical components will cancel. Similarly, a vertical field will magnetize the hysterons vertically and reduce the horizontal component to zero. This concept is the basis for choosing the following selection rules.

We will assume that if the x component of the applied field is greater than u_x, the y component is between v_y and u_y and the z component is between v_z and u_z, then the hysteron will be magnetized in the x-direction. Then, for that point in the Preisach plane

$$\mathbf{Q}(\Omega) = \mathbf{1}_{\mathbf{x}}, \tag{6.9}$$

where $u_x<h_x$, $v_y<h_y<u_y$, and $v_z<h_z<u_z$, and where the h's are operative fields. This type of selection rule will be called a *simple selection rule*. There are six such rules for three-dimensional models. They apply to each point in the Preisach hypervolume where one component of the applied field is sufficient to switch that hysteron, but the other components are not. Under these conditions, the hysteron is switched into that unique direction.

When more than one component is sufficient to switch the magnetization, *compound selection rules* govern the selection of the state vector. They are required to avoid the indeterminacy of the direction of $\mathbf{Q}$ for large fields. When they are necessary, many selections are possible. For vector models, however, the remanence is not just a function of the applied field extrema but is a function of the path taken by the applied field as it is reduced to zero. The choice of compound selection rules must take this additional dependence into account. A possible rule would be to choose the state vector to be in whichever direction the applied field has the largest excess over that component of the switching field. Thus, whichever component of the applied field last exceeds the coercivity in that direction will determine the direction of the hysteron's magnetization. This would give a continuous function for $\mathbf{Q}$ over the Preisach volume, but the derivative of $\mathbf{Q}$ is discontinuous whenever two components have the same excess over the coercivity.

(a) (b)

Figure 6.3 When both members of a pair of hysterons, at the same point in the Preisach volume, are magnetized horizontally (a), the vertical component of magnetization is zero. When both are magnetized vertically (b), the horizontal component is zero.

Furthermore, we will see later that for large fields, the irreversible magnetization does not tend to follow the applied field as it does for the following selection rules.

We now describe a better choice of compound selection rules that meet the desirable criteria we will use. When two or more components of the applied field exceed the switching fields of the hysteron, we will select the components of the state vector to be in the same ratio as the excess of the applied field's components over the respective switching field components. This would make **Q** a function with continuous derivatives over the Preisach hypervolume. To conserve space, we will summarize these rules for the two-dimensional case only, since the generalization to three dimensions is routine. The rules are summarized in Tables 6.1 and 6.2, which give the components of **Q** for these compound selection rules

Table 6.1 Values for Q_x

Q_x	$v_x > h_x$	$v_x < h_x < u_x$	$h_x > u_x$
$v_y > h_y$	$\dfrac{h_x - v_x}{\lvert h_x - v_x\rvert + \lvert h_y - v_y\rvert}$	0	$\dfrac{h_x - u_x}{\lvert h_x - u_x\rvert + \lvert h_y - v_y\rvert}$
$v_y < h_y < u_y$	-1	No change	1
$h_y > u_y$	$\dfrac{h_x - v_x}{\lvert h_x - v_x\rvert + \lvert h_y - u_y\rvert}$	0	$\dfrac{h_x - u_x}{\lvert h_x - u_x\rvert + \lvert h_y - u_y\rvert}$

Table 6.2 Values for Q_y

Q_y	$v_x > h_x$	$v_x < h_x < u_x$	$h_x > u_x$
$v_y > h_y$	$\dfrac{h_y - v_y}{\lvert h_x - v_x\rvert + \lvert h_y - v_y\rvert}$	-1	$\dfrac{h_y - v_y}{\lvert h_x - u_x\rvert + \lvert h_y - v_y\rvert}$
$v_y < h_y < u_y$	0	No change	0
$h_y > u_y$	$\dfrac{h_y - u_y}{\lvert h_x - v_x\rvert + \lvert h_y - u_y\rvert}$	1	$\dfrac{h_y - u_y}{\lvert h_x - u_x\rvert + \lvert h_y - u_y\rvert}$

as a function of the operative field. They apply to every point in the composite Preisach volume. It is noted, however, that when the applied field changes, all points are not necessarily affected and only the points affected have to be changed. Furthermore, these rules reduce to the simple selection rules when they apply.

If we assume that all the coupling between the two axes is entirely through the state vectors, then the Preisach function can be factored as

$$p(\Omega) = p_x(\Omega_x)p_y(\Omega_y)p_z(\Omega_z), \tag{6.10}$$

where

$$\Omega_x = (u_x, v_x), \quad \Omega_y = (u_y, v_y), \quad \text{and} \quad \Omega_z = (u_z, v_z). \tag{6.11}$$

Examination of Tables 6.1 and 6.2 shows that as a result of the application of the selection rules, at any point on the Preisach plane, the sum of the magnitudes of the Cartesian components of the state vector is set equal to 1; that is,

$$|Q_x| + |Q_y| + |Q_z| = 1. \tag{6.12}$$

Let us define the following two integrals:

$$I_j = \int_{\Omega_R} Q_j(\Omega)p(\Omega)\,d\Omega = \int_{\Omega_R} Q_j(\Omega_j)p_j(\Omega_j)\,d\Omega_j \quad \text{for } j = x,\ y,\ \text{or } z, \tag{6.13}$$

or

$$\mathbf{I} = \int_{\Omega_R} \mathbf{Q}p(\Omega)\,d\Omega, \tag{6.14}$$

where the Q's are computed using the selection rules as above. It is seen from (6.12) that

$$|I_x| + |I_y| + |I_z| \le 1. \tag{6.15}$$

The equality in this equation occurs only when for all points at which the Q's are not zero, all the Q_x's in I_x are of the same sign, all the Q_y's in I_y are of the same sign, and all of the Q_z's in I_z are of the same sign. For example, if the remanence is obtained by rotating a large field, then equality occurs for the entire process. We note, for example, that if I_z is zero and I_x and I_y have the same sign, so that the term $I_x + I_y$ is equal to one, then

$$I_x^2 + I_y^2 = (I_x + I_y)^2 - 2I_xI_y = 1 - 2|I_x|\cdot|1 - I_x|. \tag{6.16}$$

This equation implies that the sum of the squares of the I's is a function of I_x, hence of the direction of the magnetization. This would be true even for isotropic materials under large fields.

Thus, we cannot let

$$m_{Ix} \propto I_x,\ m_{Iy} \propto I_y \ \text{ and } \ m_{Iz} \propto I_z, \tag{6.17}$$

since the simple application of these selection rules yields neither circular remanence paths for isotropic materials nor ellipsoidal remanence paths for anisotropic materials. As the applied field is rotated from the x direction to the y direction, the normalized remanent path traces a straight line from the point (1, 0, 0) to the point (0, 1, 0). These results can easily be generalized to three dimensions. A pair of two-dimensional models [5] was proposed to correct for this limitation: the m^2 model and the m model.

6.5.2 The m^2 Model

In a possible coupled-hysteron model, the m^2 model, we compute the square of the irreversible components of the magnetization using the appropriate component of $\mathbf{Q}$. Thus, using (6.13) in two dimensions we have

$$m_{ix}^2 = I_x \ \text{ and } \ m_{iy}^2 = I_y, \tag{6.18}$$

where

$$M_{ix} = M_S S_x m_{ix} \ \text{ and } \ M_{iy} = M_S S_y m_{iy}, \tag{6.19}$$

or

$$\mathbf{M_I} = M_S\ \mathbf{m_I}\cdot\mathbf{S}, \tag{6.20}$$

where

$$\mathbf{S} = S_x\mathbf{1_x} + S_y\mathbf{1_y}. \tag{6.21}$$

If we wish the material to be ellipsoidally magnetizable, then the major remanence path must obey

$$\left(\frac{M_{ix}}{S_x}\right)^2 + \left(\frac{M_{iy}}{S_y}\right)^2 = M_S, \tag{6.22}$$

or

$$m_{ix}^2 + m_{iy}^2 = 1. \tag{6.23}$$

We see that this is indeed the case for large rotating fields, by substituting (6.15) into (6.18) with the equality sign.

The problem with this approach is that (6.18) gives only the magnitude of the components of the remanence and not their sign. The sign must be computed separately. For example, the sign of m_x could be given by a formula such as

$$\mathrm{sgn}(m_{ix}) = \mathrm{sgn}\left[\int_{\Omega_R} Q_x(\Omega)p(\Omega)\,d\Omega\right] \quad \text{and} \quad \mathrm{sgn}(m_{iy}) = \mathrm{sgn}\left[\int_{\Omega_R} Q_y(\Omega)p(\Omega)\,d\Omega\right]. \tag{6.24}$$

It is noted that in the case of a scalar applied field in the x direction, Q_x is one. However, (6.18) computes the square of the magnetization, not the magnetization directly. Thus, the vector Preisach function does not reduce simply to the scalar Preisach function. For example, an attempt to identify the Preisach function by calculating an x-directed magnetization by one starting from a negative x saturation state and applying fields only in the x direction, would not yield the same Preisach function obtained from a scalar Preisach model.

6.5.3 The Simplified Vector Model or *m* Model

A better way of coupling the two Preisach models is the m model. In this model, we use a rotational correction $R(I_x, I_y, I_z)$ to compute the normalized magnetization, and we compute the components of the normalized irreversible magnetization by means of

$$m_{Ix} = R(I_x,I_y,I_z)I_x,\ m_{Iy} = R(I_x,I_y,I_z)I_y,\ \text{and}\ m_{Iz} = R(I_x,I_y,I_z)I_z, \tag{6.25}$$

or

$$\mathbf{m_i} = R\mathbf{I} \tag{6.26}$$

where $R(I_x, I_y, I_z)$ is the rotational correction. We then compute the magnetization by substituting these expressions into

$$M_{ix} = M_S S_x m_{ix},\ M_{iy} = M_S S_y m_{iy},\ \text{and}\ M_{iz} = M_S S_z m_{iz}, \tag{6.27}$$

where the S's are the squareness of the material. Then

$$\mathbf{M_I} = M_S\ \mathbf{S}\ \mathbf{m_I}, \tag{6.28}$$

where, as a result of the choice of the coordinate system, $\mathbf{S}$ is the following matrix:

$$\mathbf{S} = \begin{bmatrix} S_x & 0 & 0 \\ 0 & S_y & 0 \\ 0 & 0 & S_z \end{bmatrix}. \tag{6.29}$$

This model is designed to handle anisotropic media by choosing different values for the S's along each of the axes, and different parameters in the basic Preisach models: the model, however, can also describe isotropic media. If we let all the S's be equal, the major remanent path will be a circle. In addition, if all the basic Preisach models have the same parameters, for any circular applied field path, all the remanent paths are circles and the model is isotropic. The model can also describe scalar processes if the applied field is along one of the principal axes. In that case, the magnetization will be along that axis. If we wish the material to be ellipsoidally magnetizable, the major remanence path must obey

$$\left(\frac{M_{ix}}{S_x}\right)^2 + \left(\frac{M_{iy}}{S_y}\right)^2 + \left(\frac{M_{iz}}{S_z}\right)^2 = M_S, \tag{6.30}$$

or

$$m_{ix}^2 + m_{iy}^2 + m_{iz}^2 = 1. \tag{6.31}$$

To obtain ellipsoidally magnetized behavior, an acceptable rotational correction could be $(I_x^2 + I_y^2 + I_z^2)^{1/2}$. For this model to reduce to the scalar model, it is desirable for this correction to approach unity under these conditions. Thus, for the m model, we will use the rotational correction given by

$$R(I_x, I_y, I_y) = \frac{|I_x| + |I_y| + |I_z|}{\sqrt{(I_x^2 + I_y^2 + I_z^2)}}. \tag{6.32}$$

In the two-dimensional case $I_z = 0$. Then if $I_x = 0$, we have $R = 1$, $m_x = 0$, and $m_y = I_y$. Similarly, when $I_y = 0$, again $R = 1$, but now $m_y = 0$, and $m_x = I_x$. It can be shown that for all values of I_x or I_y

$$1 \le R \le \sqrt{3}. \tag{6.33}$$

Thus, if we apply a large field along either the x or the y axis, the corresponding Q goes to one for all points on the Preisach volume, and the other Q goes to zero. Thus, for processes in which the applied field always lies along the x axis, after the application of a very large field, the rotational correction will be unity and the irreversible magnetization will be given by

$$m_{ix} = \int_{u_x > v_x} Q_x(\Omega_x) p_x(\Omega_x)\, d\Omega_x, \tag{6.34}$$

where

$$Q_x(\Omega_x) = \int_{\Omega_R} \mathbf{1}_x \cdot \mathbf{Q}(\Omega)\, d\Omega_y. \tag{6.35}$$

So, for these processes, the m model reduces to the ordinary scalar processes. Similarly, processes along either the y or z axes also reduce to ordinary scalar processes. Thus, the model can have noncongruency and exhibit aftereffect and accommodation. Also for incremental changes in the applied field only a small region of the Preisach volume will change, so the differential equation approach to magnetization changes can be very effective. Therefore the models can be identified individually in the same manner used for scalar processes.

If we compute the magnitude of the magnetization for the remanence due to a large field in any direction, since

$$|I_x| + |I_y| + |I_z| = 1, \tag{6.36}$$

and the rotational correction is still $\sqrt{(I_x^2 + I_y^2 + I_z^2)}$, we have

$$(RI_x)^2 + (RI_y)^2 + (RI_z)^2 = 1. \tag{6.37}$$

Thus, we see that

$$m_{ix}^2 + m_{iy}^2 + m_{iz}^2 = 1. \tag{6.38}$$

This states that the normalized major remanent path lies on a sphere, and thus, the major remanent path itself lies on an ellipsoid unless all the S's are equal. If more complex paths are desired, additional rotational corrections can be added.

In particular for isotropic media, for large h the selection rules require that

$$Q_j = \frac{h_j}{|h_x| + |h_y| + |h_z|}, \quad \text{where } j = x, y, \text{ and } z. \tag{6.39}$$

Since the field is large,

$$I_j = \iint_{\Omega_R} Q_j \, p_j(\Omega) \, d\Omega = Q_j, \quad \text{where } j = x, y, \text{ and } z. \tag{6.40}$$

From (6.27) we see that

$$m_{ij} = \frac{Q_j}{\sqrt{Q_x^2 + Q_y^2 + Q_z^2}}, \quad \text{where } j = x, y, \text{ and } z. \tag{6.41}$$

Thus, again

$$m_{ix}^2 + m_{iy}^2 + m_{ix}^2 = 1. \tag{6.42}$$

Furthermore, for an applied field rotating in the xy plane,

$$\frac{Q_x}{Q_y} = \frac{h_x}{h_y} = \frac{m_{ix}}{m_{iy}}. \tag{6.43}$$

Thus, the magnetization will be aligned with the applied field and will have constant magnitude.

If the individual scalar process is modeled with accommodation, aftereffect, and state-dependent reversible magnetization, and is a moving model, the resulting vector model will have all these properties. In this case, for an applied rotating field, the magnetization path will be an elliptical helix whose pitch decreases with each rotation until finally it reaches an elliptical limit cycle, as shown in Fig. 6.4.

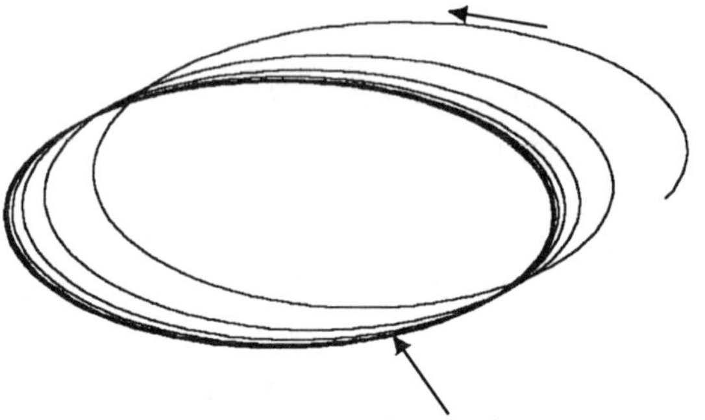

Figure 6.4 Magnetization path of an accommodating anisotropic medium due to a rotating applied field.

For isotropic media, the Preisach models along the x and y axes are identical, so only one identification is necessary. For anisotropic media, the parameters of the three models will be different, especially the mean critical fields and the squarenesses. Then, for large fields, the irreversible component of the magnetization is in the same direction as the applied field only when the applied field lies along the principal axes. In general, the magnetization will lie closer to the easy axis. For smaller fields, the magnetization will also lag behind the applied field, and the aspect ratio of these paths can be different from that of the major path.

So far we have computed the irreversible component of the magnetization. The reversible component of the magnetization is also a vector and can be computed by

$$M_{Rj} = (1 - S_j) M_S |I_j| R(I_x, I_y, I_z)[a_{j+} f(H_j) - a_{j-} f(-H_j)], \tag{6.44}$$

where $j = x, y, z$, or the normalized reversible component of the magnetization can be computed by

$$m_{Rj} = |I_j| R(I_x, I_y, I_z)[a_{j+} f(H_j) - a_{j-} f(-H_j)], \tag{6.45}$$

where again $j = x, y, z$, and the I's are the same in the case of the irreversible component. This can be implemented with either a state-independent, magnetization-dependent, or state-dependent reversible magnetization as in the case of the scalar models. It is seen that if I_y and I_z are zero, then $|I_x| R(I_x, I_y, I_z)$ is one, and so m_x is the ordinary scalar reversible component of magnetization, and the other m_r's are zero.

In any of these cases, if the field has a sufficiently strong x component, then

$$|a_{x+} f(H_x) - a_{x-} f(-H_x)| = 1. \tag{6.46}$$

Therefore, for sufficiently strong fields, the normalized reversible magnetization satisfies

$$m_{Rx}^2 + m_{Ry}^2 + m_{Rz}^2 = 1, \tag{6.47}$$

and

$$\left(\frac{M_{Rx}}{(1-S_x)M_S}\right)^2 + \left(\frac{M_{Ry}}{(1-S_y)M_S}\right)^2 + \left(\frac{M_{Rz}}{(1-S_z)M_S}\right)^2 = 1. \tag{6.48}$$

Thus, for both isotropic and anisotropic media in the presence of large fields, the normalized reversible component of magnetization has a constant magnitude, and the reversible magnetization traces out an ellipse. The magnitude of the reversible magnetization, therefore, traces out an ellipse whose major axis is the easy axis and whose minor axis is the hard axis, as shown in Fig. 6.5. It can be shown using (6.37) that

$$\begin{aligned} M_x^2 + M_y^2 + M_z^2 &= (M_{Ix}+M_{Rx})^2 + (M_{Iy}+M_{Ry})^2 + (M_{Iz}+M_{Rz})^2 \\ &= M_S^2[(RI_x)^2 + (RI_y)^2 + (RI_z)^2] = M_S^2. \end{aligned} \tag{6.49}$$

Thus, the magnitude of the magnetization is a constant equal to M_S in the direction of the applied field. Since for anisotropic media the irreversible magnetization lies between the applied field and the easy axis, the reversible magnetization lies between the applied field and the hard axis, as shown in Fig. 6.6.

For this rotational correction, the model is, in general, elliptically magnetizable and has the saturation property. When magnetized along either the easy axis or the hard axis, the model reduces properly to the scalar model, and the simplified model can be computed directly. This simplifies the identification of the parameters. It is noted that when the applied field is not along a principal axis, none of the models reduce to simple Preisach models because the magnetization is not in the same

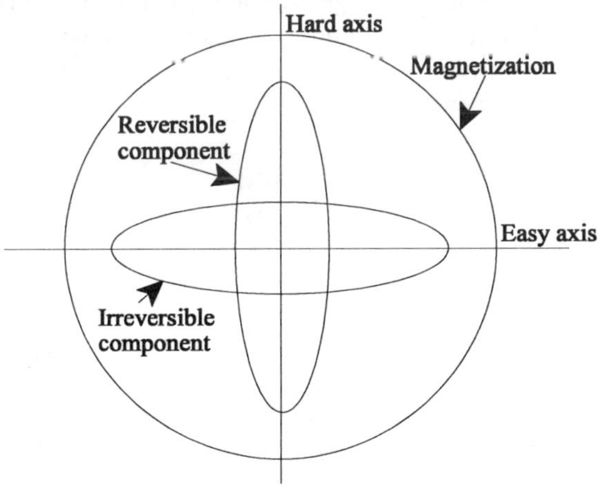

Figure 6.5 Magnetization loci for a large rotating field.

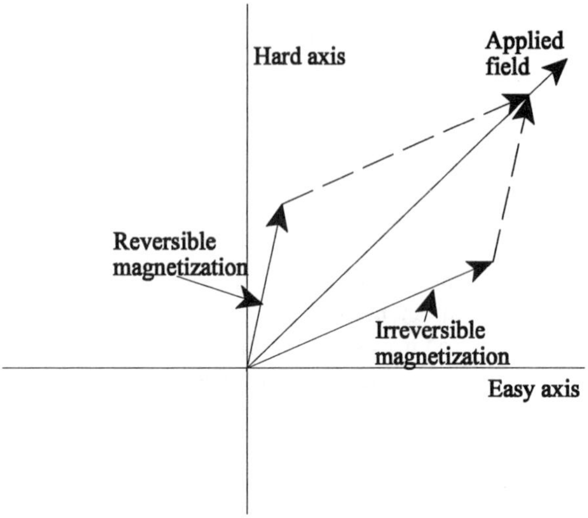

Figure 6.6 Vector decomposition of magnetization.

direction as the applied field. Since the model involves only the computation of Preisach models along the principal axes, like the scalar Preisach model, it is computationally efficient.

6.6 LOSS PROPERTIES

In the case of the Mayergoyz model and the coupled Preisach model, one is computing the irreversible component of the magnetization, while in the pseudoparticle model one computes the total magnetization. Thus, one must add a reversible component to the first two categories of models. For isotropic media, all the models predict that for large applied rotational fields, the computed magnetization will be in the same direction as the applied field. Any reversible magnetization will also be in that direction. The energy the field supplies to a magnetic medium is given by

$$\frac{dw}{dt} = \mathbf{H} \cdot \frac{\mathbf{dM}}{\mathbf{dt}}. \tag{6.50}$$

Since the magnitude of the magnetization is constant, its time rate of change must be perpendicular to it, and thus, no energy will then be supplied to the material. The stored energy in the reversible component of the magnetization does not change, because the magnitude of the vector remains the same.

For anisotropic media, the total magnetization will still be in the direction of an applied field if it is large enough; however, the models that compute the irreversible component of the magnetization compute a component that lags behind the rotating field. Thus, they would compute an energy supplied to the medium. For the total magnetization to be in phase with the applied field, the irreversible component must then lead the applied field. The amount of lead depends on the irreversible state; thus, the reversible magnetization must be state dependent. Furthermore, it would compute energy given up by the medium which is equal to that supplied to the irreversible component of the magnetization. Thus, the net energy supplied to the medium in this case is also zero. For smaller fields, not only does the irreversible magnetization start lagging behind, but also the lead of the reversible component decreases. Hence, there will be hysteresis loss in the material.

6.7 CONCLUSIONS

Vector hysteresis models must obey all the physical realizability conditions of scalar models. These limits put certain constraints on the parameters of a model. These constraints include the conditions that the magnetization cannot exceed saturation, and that the energy dissipated by the material, for any change in applied field, must be positive. The latter constraint includes the crossover condition [6] which prevents minor loops from being traversed in the clockwise direction. In addition, vector models should be able to calculate magnetizations that do not exceed saturation and also correctly calculate the energy loss for large rotating fields. For rotating fields, these losses for most materials must eventually decrease as the amplitude of an applied rotating field increases, but for oscillating fields, they must saturate as the amplitude of an applied field increases.

Many vector models have been proposed that have the correct rotational properties and reduce to scalar Preisach models under the appropriate conditions. Of these, the m model is the most computationally efficient. It is also the easiest one to correct for observed deviations from the classical Preisach model, such as accommodation and aftereffect.

REFERENCES

[1] I. D. Mayergoyz, *Mathematical Models of Hysteresis*, Springer-Verlag: New York, 1991.

[2] J. Oti and E. Della Torre, "A vector moving model of both reversible and irreversible magnetizing processes," *J. Appl. Phys.*, **67(9)**, May 1990, pp. 5364–5366.

[3] J. Oti and E. Della Torre, "A vector moving model of non-aligned particulate media," *IEEE Trans. Magn.*, **MAG-26**, September 1990, pp. 2116–2118.

[4] E. Della Torre and F. Vajda, "Vector hysteresis modeling for anisotropic recording media," *IEEE Trans. Magn.*, **MAG-32**, May 1996, pp. 1116–1119.

[5] F. Vajda and E. Della Torre, "A vector moving hysteresis model with accommodation," *J. Magn. Magn. Mater.*, **155**, 1996, pp. 25–27.

E. Della Torre and F. Vajda, "Vector hysteresis modeling for anisotropic recording media," *IEEE Trans. Magn.*, **MAG-32**, May 1996, pp.

1116–1119.

[6] F. Vajda and E. Della Torre, "Characteristics of magnetic media models," *IEEE Trans. Magn.*, **MAG-28,** September 1992, pp. 2611–2613.

CHAPTER 7

PREISACH APPLICATIONS

7.1 INTRODUCTION

This chapter introduces several indirect applications of the Preisach model. One application deals with modifications to include dynamic effects. Another explains how magnetostriction can be introduced into the Preisach formalism. These applications involve coupling to other fields, such as eddy current fields or stress fields induced by the material's magnetization. They are presented to indicate the generality of Preisach modeling.

7.2 DYNAMIC EFFECTS

In Chapter 5, we discussed aftereffect, which is principally a long time-constant effect. We will now discuss short time-constant dynamic effects. The two principal short time-constant dynamic effects are: *eddy currents* in conductors and inertial effects, such as *gyromagnetism*. Eddy currents are induced in conductors whenever the field changes. In the case of magnetic materials the change in magnetization can in turn induce eddy currents. Eddy currents have the effect of shielding the interior of the material from changes in the applied field. Thus, there is a strong spatial interaction involved in the computation of the material's behavior.

In nonconductors, the principal dynamic effects are gyromagnetic; that is, when a magnetic moment is placed in a magnetic field, its moment precesses about the field and eventually aligns itself with the field by dissipating some of its energy, since the aligned state is lower in energy. Thus, as we saw in Chapter 1, precession causes a domain wall to move with finite mobility, and causes the phenomenon of ferromagnetic resonance. In other geometries, many additional complex effects can be observed, such as nonreciprocity. These effects are beyond the scope of this book and are not discussed further. The example of dynamic effects that we will discuss in the next sections are associated with eddy currents and reversal times.

7.3 EDDY CURRENTS

To understand the effects of eddy currents in magnetic materials, we consider first a simplified model, seen in Fig. 7.1, in which a tape of magnetic material is wound into a thin toroid whose inner diameter is almost equal to its outer diameter. We further wrap a conducting wire around the toroid, carrying a current I, to produce an almost uniform field inside the tape. We assume that the tape is made of a uniform ferromagnetic material, and is rectangular in cross section, as shown in Fig. 7.2. Furthermore, we assume that the coercivity is uniform throughout the material, that the tape is uniformly magnetized when saturated, and that the magnetization changes by nucleating a domain wall at each surface that propagates inward and reverses the magnetization of each region that it passes.

The motion of the wall is retarded by eddy currents. They have the effect of shielding the interior of the tape from the applied field. For this geometry, we can calculate all the fields if we neglect the effect of the ends. Then the eddy currents are uniform in the region between the surface and the domain wall, and zero inside the domain wall. Their effect is to reduce the applied field to the coercive field at the domain wall so that the wall can begin to move. The dynamic behavior of the magnetizing process is determined by balancing the wall's velocity with the effect of the eddy currents. If the wall velocity is too large, then the field at the wall will fall below the coercivity and the wall cannot move. If the velocity is too small, then

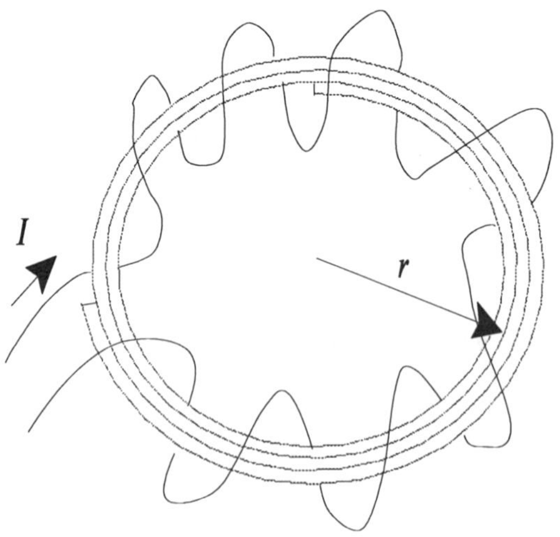

Figure 7.1 Toroid used to illustrate eddy current effects.

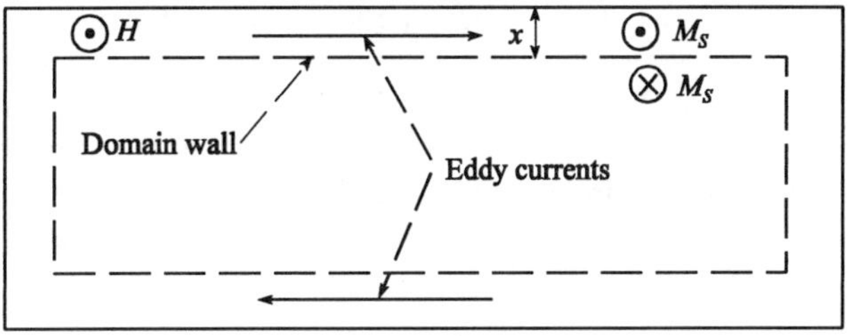

Figure 7.2 Cross section of a tape.

there will be insufficient shielding, and the wall will be accelerated.

With this geometry the eddy current density is uniform, so the total eddy current is given by I, the eddy current density by J, and the distance that the wall is from the surface by x. Therefore, the field at the wall H_W, equal to the applied field H less the effect of the eddy currents, is given by

$$H_W = H - Jx. \tag{7.1}$$

The applied field is given by

$$H = \frac{NI}{r}, \tag{7.2}$$

where N is the number of turns in the magnetizing coil and r is the radius of a given tape element. Thus each turn of the tape experiences a slightly different field. The eddy currents are determined by Ohm's law; that is,

$$J = \sigma E, \tag{7.3}$$

where E is the field induced by the eddy currents. This field is the negative of the rate of change of magnetic flux divided by the path length. If the tape thickness is s, the rate of change of magnetic flux per unit length is given by

$$E = M_S \frac{dx}{dt} \quad \text{for} \quad x \le s/2. \tag{7.4}$$

The total shielding current is computed by reducing the applied field to the coercivity at the domain wall. Therefore, we have

$$I = \sigma M_S x \frac{dx}{dt} = H - H_C. \tag{7.5}$$

This equation could be solved to give us the net magnetization $M(s - 2x)$ as a function of the applied field.

This model would assume that every time the applied field changes sign, a new domain wall starts propagating inward from the surface. Unfortunately, the behavior of a real material is much more complicated. The nucleation of a domain wall requires fields much higher than those required to propagate it. Furthermore, the coercive field is a random variable of the position. Thus, the domain wall does not propagate inward as a plane, but becomes distorted and may even break up into many sections.

An alternate approach is a nongeometric one in which average magnetization is computed without worrying about how it is distributed in the material. Bertotti suggested [1] that each point in the Preisach plane has a state, Q, that varies continuously between -1 and 1 as a function of time. He then computes the magnetization as a function of time by

$$M(t) = SM_S \iint_{u>v} p(u,v)\,Q(u,v,t)\,du\,dv. \tag{7.6}$$

If at a particular point in the Preisach plane, the applied field is greater than u, then for that point the state function will vary according to

$$\frac{\partial Q}{\partial t} = \begin{cases} k\cdot[h(t)-u], & \text{when} \quad h > u \\ k\cdot[h(t)-v], & \text{when} \quad h < v, \end{cases} \tag{7.7}$$

where k is an unknown parameter. This parameter varies inversely with the conductivity of the material and would be infinite if the material had zero conductivity. In this case, the state function would change instantaneously whenever the field exceeded u, as in the case of the classical Preisach model. Thus, this calculation correctly reduces to the classical model for insulators.

This model predicts a hysteresis loss as a function of magnetizing frequency that can be described by

$$w = c_1 + c_2\sqrt{\omega}, \tag{7.8}$$

where w is the hysteresis loss per cycle, ω is the frequency of the applied field, and c_1 and c_2 are monotonic increasing functions of the peak of the applied sine wave magnetic field. The latter two constants are a function of the material and the geometry. This is consistent with measurements.

7.4 FREQUENCY RESPONSE OF THE RECORDING PROCESS

The frequency response of the recording process is determined principally by the recorded wavelength. Thus, if the media speed past the recording head is increased, the recorded wavelength is increased for a given frequency signal. Thus, neglecting the effect of the circuit parameters in the head, doubling the speed effectively doubles the frequency response of the process. Of course the reactances associated with the head windings and the ability of the media to respond to the applied fields will ultimately limit the ability of the medium to respond to the signal.

Two factors control the frequency response of the recording process. The first is due to the inability to localize the magnetic field. Thus, even if one uses a ring-type recording head with zero gap width, the magnetic field is not very well localized, as shown in Fig. 7.3. In particular, one gap length away from the head, the perpendicular component of the magnetic field eventually decreases as the reciprocal of the distance from the gap, and the longitudinal field eventually decreases as the square of the reciprocal of the distance from the gap. It is seen that a vector model is necessary to analyze properly the recording process. Even though the strongest component of the magnetic field is longitudinal, there is a sizable perpendicular component. Furthermore, the perpendicular component eventually dominates the magnetizing process.

The second factor that controls the recording process is related to the characteristics of the recording media. Even if the standard deviation of the switching field is zero, a transition would have finite width because of the spread of the head field. We see that a nonzero switching field distribution also affects the frequency response. This is because it takes the vulnerable part of the medium a finite time to pass the region of the recording head where the field is of the order of the coercivity.

For a step function in the applied field and zero switching field distribution, the transition occurs at the place where the applied field has decreased to the coercivity of the material. Therefore, if the amplitude of the applied field is changed, the location of the transition will move to a position that satisfies this criterion. If the switching field distribution is not zero, the transition will have a finite width. The hysterons with the smallest switching fields will be written furthest downstream from the gap, and the hysterons with the largest switching

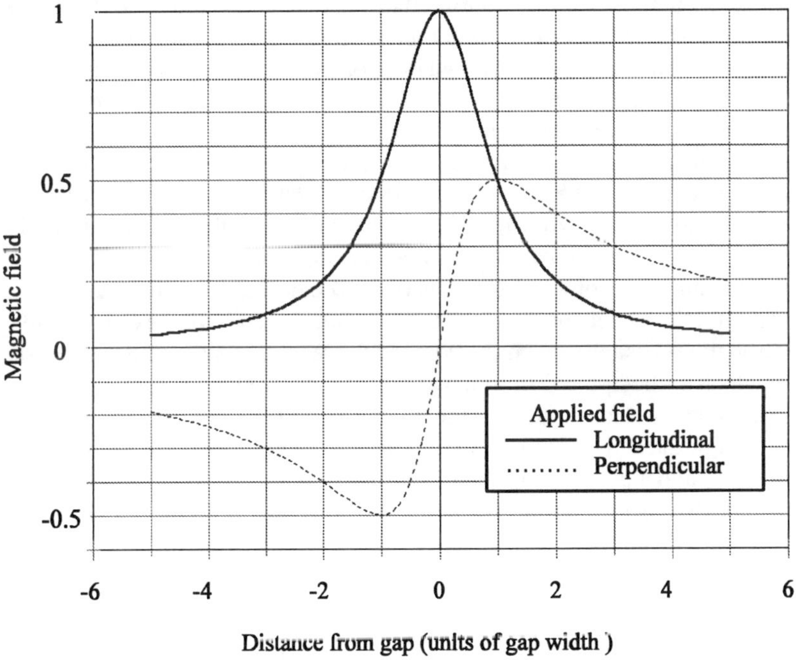

Figure 7.3 Field due to a ring-type head.

fields, which are affected by this head, will be written closest to the gap. This effect, for *ac*-bias recording, is discussed elsewhere [2].

7.5 PULSED BEHAVIOR

Accommodation has been observed in particulate recording media [3,4], under pulsed conditions. It appears that the source of this effect may be the statistical stability of the Preisach model. We now examine how this variation can be explained by the Preisach accommodation model discussed in Chapter 5. When a field is applied, the magnetization normally will change, causing all the hysterons to move within the Preisach plane as well. Normally, hysterons whose positive operative switching fields are less than an applied *dc* field will switch to their positive state. While the field is reduced to zero, for fourth-quadrant media, normally no further switching occurs; however, the final magnetic state of the system will be different if the pulses are long enough to nucleate a magnetization reversal, but not long enough for the reversal to take place.

The experiment described by Flanders et al. [3] involves the change in the final remanence caused by a pulse when it precedes a larger pulse. There are two sources for this difference in the model described below: the motion of the distribution as a whole, which is described by the moving model, and the motion of hysterons within the distribution, which is described by the Preisach accommodation model as the source of the accommodation.

It has been known for some time that in soft materials the field required to nucleate a domain wall is much larger than that required to propagate it. Numerical micromagnetic studies [5] have shown that for hysterons used in recording media, the field required to nucleate a reversal H_N is also much larger than that required to propagate the reversal throughout the hysteron.

The simulation of the dynamics in this process is based on two characteristic times associated with the reversal process: the nucleation time and the actual reversal time. Nucleation time t_N is the length of time that the nucleation field must be applied in order for the magnetization reversal to nucleate; reversal time t_R is the total time required to complete that reversal. The nucleation time decreases if the field applied is increased beyond the minimum field, but even the longest nucleation time is usually much shorter than the reversal time. When a reversal has been nucleated, the applied field may be drastically reduced, sometimes even to negative values, without affecting the completion of the reversal process.

Although each hysteron in a medium can have a different applied-field-dependent t_N and a different applied-field-dependent t_R, to simplify the model we will assume that these times are the same and constant for all hysterons. The field dependence of t_R is not a serious source of error, since in the experiment described, the field will always be zero during reversal. We assume that a hysteron subjected to a field pulse, whose strength is H_N and whose duration is t_P, will switch if $t_P > t_N$, and will not switch if $t_P < t_N$. Furthermore, if $t_N < t_P < t_R$, the hysteron will reverse, but it will complete its reversal after the pulse has ended.

7.5.1 Dynamic Accommodation Model

The source of accommodation in the model discussed in Chapter 5 is the motion of hysterons in the Preisach plane whenever the magnetization changes. Unlike the hysterons of the classical Preisach model, in its new position a hysteron might find itself with a different magnetization from nearby hysterons — perhaps because it acquired its magnetization at its old position, and during its motion in the operative plane did not experience a field large enough to reverse it. In the accommodation model, it was assumed that the field was applied until the magnetization had achieved steady state. For short pulses it is probable that a hysteron's position will start to change after the pulse has ended. It will then have a different remanence from that which it would have had if the field had been kept constant until steady state had been achieved.

The Preisach plane is divided into six regions, as shown in Fig. 7.4, for a medium whose history is suggested by the staircase line. At zero field, the magnetization in region I is always kept positive, and the magnetization in region VI is always kept negative. In regions II and III the magnetization is essentially positive owing to its history, but may become diluted as a result of accommodation. Similarly, regions IV and V are essentially magnetized negatively. Application of H_A nucleates reversals in region V, which starts the motion of the hysterons in the Preisach plane. Subsequent application of a field pulse H_A will nucleate any negative hysterons that have moved into regions II and V, leading to accommodation. Table 7.1 compares the magnetization state of a region before the application of a field with that computed by Preisach models and with that computed by this model after the application of a pulse.

In Table 7.1, "Same" indicates that the hysteron remains in the state defined in the region it came from, which may be different from that computed by the classical Preisach model. It is seen that for long pulses, the Preisach accommodation model dilutes regions III and IV, while short pulses dilute regions

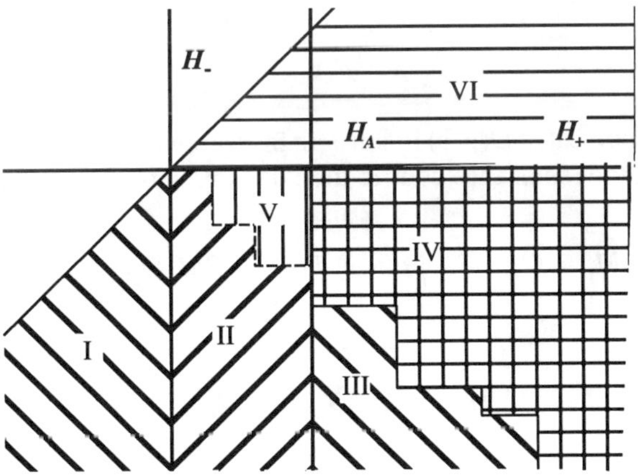

Figure 7.4 Division of the Preisach plane into six behavioral regions.

II and V as well. Thus, there is a change in magnetization at the conclusion of each pulse. After many pulses, it is expected that the Preisach accommodation model will asymptotically approach the same equilibrium magnetization for either short or long pulses, but this magnetization will be different from that computed by the classical Preisach model. It was shown in Chapter 5 that the amount of dilution depends on the average magnetization and the change in magnetization.

Table 7.1 Hysteron Magnetization State

Region	Previous State	Preisach Models	Pulse $t_p > t_R$	Pulse $t_p < t_R$
I	+	+	+	+
II	+	+	+	same
III	+	+	same	same
IV	-	-	same	same
V	-	+	+	same
VI	-	-	-	-

We will assume that the medium is a single-quadrant medium, so that $P(u,v)$ is zero if either u is negative or v is positive. The state variable Q can take any value from −1 to 1, to account for the dilution of the region due to the motion of hysterons in the Preisach plane. If we define the component of the Preisach function due to region j (j = I, II, III, etc.) by

$$p_j = \int_j p(u,v)\,du\,dv, \tag{7.9}$$

then the normalization is

$$\sum_j p_j = \int_{u>v} p(u,v)\,du\,dv = 1. \tag{7.10}$$

We will also assume that $Q(u, v)$ is constant in any region of the Preisach plane and define M_j to be the remanence contribution due to region j in the operative plane; then

$$M_j = Q_j\, SM_S \int_j p(u,v)\,du\,dv = SM_S Q_j p_j. \tag{7.11}$$

Thus,

$$M_i = SM_S \sum_j Q_j p_j. \tag{7.12}$$

If a field H_1 is applied to a medium that is negatively saturated, the Preisach plane is divided as shown in Fig. 7.5, where h_1 is the operative field, $H_1 + \alpha M$, and α is the moving parameter. For pulses such that $t_P > t_R$, the state variable Q_I, associated with region I, will be +1; however, if $t_N < t_P < t_R$, it will be diluted to a smaller, but still positive, value. The subsequent application of field H_2 will increase the value of Q in region II from –1 to a maximum of +1, if it is held for a sufficiently long time.

We will define $M(H_2)$ to be the remanence after a negatively saturated medium has been subjected to a field pulse, H_2, and we will define $M(H_1, H_2)$ to be the remanence after the same negatively saturated medium has been first subjected to a field pulse H_1, followed by a field H_2, where $H_1 < H_2$. The experiment described earlier [3] compares $M(H_2)$ with $M(H_1,H_2)$; these authors found that fewer of the hysterons switched in the second case, especially when H_2 is the order of the coercivity. It is noted that $M(H_2) = M(0, H_2)$.

In the model presented here, two effects account for this difference. The first is due to the difference in operative fields. For a single pulse, the operative field is $H_2 + \alpha\ [M_r(H_2) - S\ M_S]$, where $M_r(H_2)$ is the reversible component of magnetization when H_2 is applied. When two pulses are applied, the operative field at the second pulse is $H_2 + \alpha[M_1 + M_r(H_2)]$, where M_1 is the magnetization due to the first pulse. This field operative is more positive than $-S\ M_S$. For positive α, $M(H_1, H_2)$ is a monotonically increasing function of H_1; for negative α it is a monotonically decreasing function. In particular, if there is no reversible magnetization and the Preisach function is Gaussian, the remanence is proportional to the error function of H_1.

The second effect is due to accommodation [3]. The state variable in region III, after the application of H_1, is given by

$$Q_{\mathrm{III}1} = Q_{\mathrm{III}0} + \beta <M> \Delta M, \tag{7.13}$$

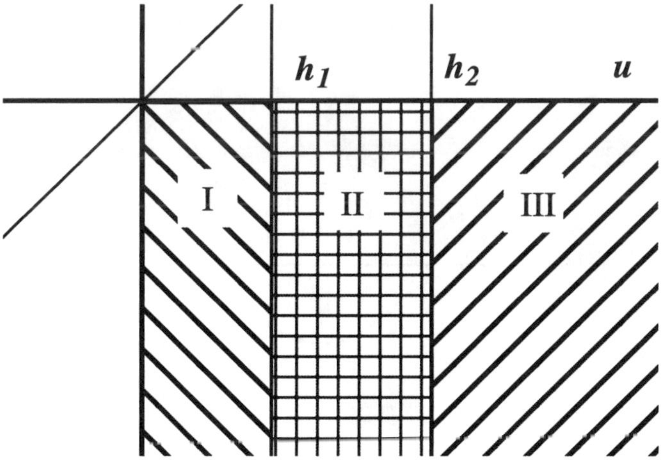

Figure 7.5 Division of the operative plane when fields h_1 and h_2 are applied.

where $Q_{III\,0}$ and $Q_{III\,1}$ are the initial and final state variables, respectively, β is an accommodation constant that determines the fraction of hysterons at a point on the Preisach plane that come from other regions, $<M>$ is the steady-state average remanence, and ΔM is the total change in magnetization. Since ΔM is equal to $2p_I$ and $<M>$ is equal to $SM_S\,(p_I - p_{II} - p_{III})$, we see that

$$Q_{III1} = -1 + \gamma P_I, \tag{7.14}$$

where $\gamma = 4\,\beta SM_S$. Then, from (7.12), after the application of the second pulse, the resulting remanence is

$$M_2(H_1,H_2) = p_I + p_{II} - p_{III} + \nu p_{III}\, p_I^2. \tag{7.15}$$

In the case of a single pulse equal to the coercivity, we have $p_{II} = p_{III} = 0.5$, and $M(0, H_2) = 0$. For a double pulse, we see that $M(H_1, H_2) = 1 - 2p_{III} + \nu p^2_{III}$, and therefore, $M(H_1, H_c) = \nu p^2_{III}$. For example, if H_1 is chosen such that $p_I = 0.25$, then for H_2 still at the coercivity, $p_{II} = 0.25$, $p_{III} = 0.5$, and thus, $M(H_1, H_2) = \nu/32$.

To reproduce the results in [3], one could assume that α is negative and that the two effects described above are roughly equal when $H_1 = H_2$. A calculation of the remanence difference, $\Delta = M(H_1, H_2) - M(0, H_2)$, as a function of H_1/H_2, for high squareness media is shown in Fig. 7.6. This is similar to the result obtained experimentally in that paper [3]. A quantitative analysis of this effect would require the identification of a complete set of the model parameters for a given medium.

Modeling the overwrite process in very high frequency recording requires a simple model that calculates the variation of the remanence with pulse height and width of the applied field. The model we present here is the DOK model, a moving model with magnetization-dependent locally reversible magnetization [6], to which we have added accommodation effects [3]. This extension of our results [7], assumes that once the critical field for a hysteron has been reached, its magnetization will start to change only after a nucleation time, t_N, whereupon, it

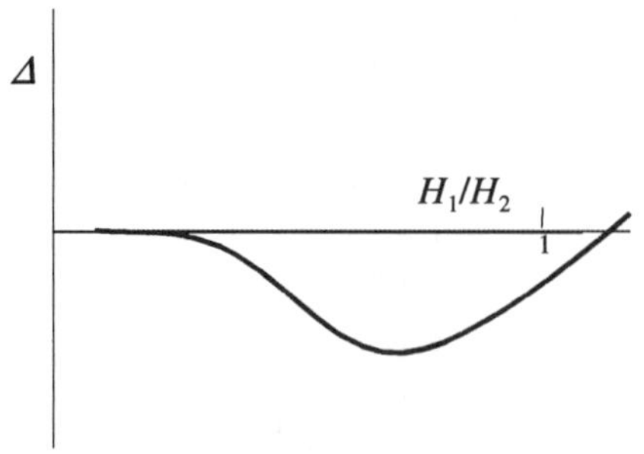

Figure 7.6 Ratio of remanences as a function of H_1/H_2.

will rotate at such a rate that its magnetization varies linearly from state to state in a time, t_R, even if the applied field is then removed.

We will compute the variation of the remanence of a medium that has been initially saturated negatively (*down*) after the application of two positive pulses (*up*) of various heights and lengths and compare these results with the measurements of Doyle et al. [3]. Before further refinement of the model is undertaken, one must identify the medium's parameters through careful analysis and compare the model results quantitatively with experiments.

The irreversible component is obtained by integrating the product of the Preisach function $P(u, v)$ and the state function $Q(u, v)$ where u and v are the "up-" and "down-" switching fields, respectively. Thus,

$$m_i = \int_{u>v} Q(u,v)p(u,v)\,du\,dv = \sum_j Q_j p_j . \tag{7.16}$$

The state function is either +1 or –1 for the classical Preisach model, but in the accommodation model, because of dilution, it can take an intermediate value. There are three ranges of applied field to be considered: If the applied field is larger than the value of u in a region, Q is set to +1; if the applied field is smaller than the value of v in a region, Q is set to –1; otherwise, Q is unaffected by that field.

Accommodation occurs when the magnetization changes and the interaction field changes at all hysterons. Thus, the positions of hysterons in the operative plane change. Therefore, the value of Q in an unaffected region is modified by hysterons coming into that region from other parts of the plane, carrying with them their original magnetization. As in Chapter 5, we will assume that the value of Q in such a region is given by

$$Q = (1 - \beta\Delta m)Q' + \beta|\Delta m|\langle m\rangle , \tag{7.17}$$

where Q' is the old value of Q, β is the accommodation constant, Δm is the change in normalized magnetization, and $\langle m\rangle$ is the average normalized magnetization.

In this model we will use the DOK characterization of the locally reversible component of the magnetization, so that

$$m_r = \frac{m_i + 1}{2} f(H) + \frac{m_i - 1}{2} f(-H), \tag{7.18}$$

where $f(H)$ is the variation of the reversible magnetization when a hysteron is in the "up" state. In the following simulation, we will use the following function for $f(h)$:

$$f(h) = 1 - \exp\left(-\frac{\chi H}{M_S}\right). \tag{7.19}$$

Although this is a monotonic increasing function, its slope is a monotonic decreasing function. Since $f(H)$ is zero if H is zero, and $f(H)$ is always greater than $f(-H)$, if H is held constant, then from (7.18), as m_i increases, m_r will decrease.

Defining the region R_1 to be the physical region of the operative plane to the left of the line $h = h_1$ is convenient. If we assume that the Preisach function is Gaussian, then it has been shown [8] that p_1 is given by

$$p_1 = \frac{1}{2}\left[\operatorname{erf}\left(\frac{h_1 - h_c}{\sigma}\right) + 1\right], \tag{7.20}$$

where erf is the error function. Defining the remainder of the physical region of the operative plane to be R_2, when an "up" field of strength H is applied to a medium that is in the "down" state, we have

$$m_i = p_1 + Q_2 p_2, \tag{7.21}$$

where Q_2 would be -1 if there were no accommodation, but now is given by (7.17), where Q' is -1.

When an "up" pulse whose time duration is greater than t_N is applied to a hysteron, we will assume that the variation of its moment with time is given by

$$g(t) = \begin{cases} -1 & \text{if} \quad t < t_N \\ \dfrac{t - (t_N + t_R/2)}{t_R/2} & \text{if} \quad t_N < t < t_N + t_R \\ 1 & \text{if} \quad t_N + t_R < t\,. \end{cases} \tag{7.22}$$

Therefore, m_i as a function of time is given by

$$m_i(t) = p_1\, g(t) + \frac{p_2}{2}\,[Q_2 + 1 + g(t)(Q_2 - 1)], \tag{7.23}$$

where the state of region 2 varies from $Q = -1$ to $Q = Q_2$.

7.5.2 Single-Pulse Simulation

We will now assume that the medium is saturated "down" and that at $t = 0$ an "up" pulse whose strength is H_1 and whose duration is t_{D1} is applied. As long as t is less than t_N, nothing happens. After that, the magnetization will start to change linearly; however, for positive α as the magnetization changes, the operative field will increase, thereby increasing the slope. If the duration of the pulse is long enough to permit all the hysterons that are going to switch to complete their switching, the system will be in equilibrium at the conclusion of the pulse.

Although the applied field is constant during the pulse, the operative field h varies with the magnetization. The irreversible magnetization varies according to (7.23) and, although $f(H)$ and $f(-H)$ remain constant, the reversible magnetization varies because the state changes according to (7.18). Thus, the operative field is given by

$$h_1 = H_1 + \alpha M_S[m_i(t) + m_r(t)]. \tag{7.24}$$

For positive pulses, $m_i(t)$ will increase, which in turn causes $m_r(t)$ to decrease. Thus, these two magnetization changes are in opposite directions. To solve for the magnetization, one must substitute this operative field into (7.20) to compute m_i using (7.21) and obtaining m_r from (7.18). Since these equations are implicit in m_i, they have to be solved iteratively.

Figure 7.7 illustrates the variation of the magnetization with time for a pulse whose duration, 6 arbitrary units, is less than the reversal time of a hysteron. When the pulse is applied, m_r immediately responds. The change in m_i after the nucleation time of 3 units, causes m_r to decrease, since it is state dependent. At the conclusion of the pulse, m_r immediately decreases to zero; however, the model assumes that m_i and m both continue to change until t_R. The total magnetization, the solid line, is simply the sum of these two components.

The calculated variation in the remanence is a step function of the pulse width, as shown in Fig. 7.8. There is no change in the remanence until the nucleation time is reached. After that, the remanence changes whether the pulse is there or not. When the pulse is finished, the change in magnetization will cause the change in location of hysterons in the Preisach plane that is the cause of accommodation; however, the motion of hysterons in the plane cannot change the remanence unless they encounter an applied field, which is now zero, greater than their switching field. The pulse width dependence changes only the initial conditions for the application of a second pulse.

The step function behavior is due to the model's assumption that once its critical field has been exceeded, a hysteron will continue to reverse, even if the applied field is turned off. If one modifies this behavior to that of reversing only a fraction of the hysterons depending on the fraction of the magnetization change that has occurred, then one would get a ramp increase in the remanence with pulse

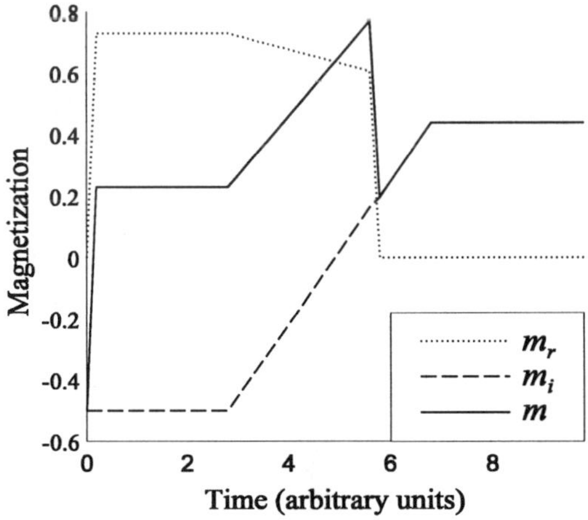

Figure 7.7 Variation of the total magnetization and its components when a single pulse is applied.

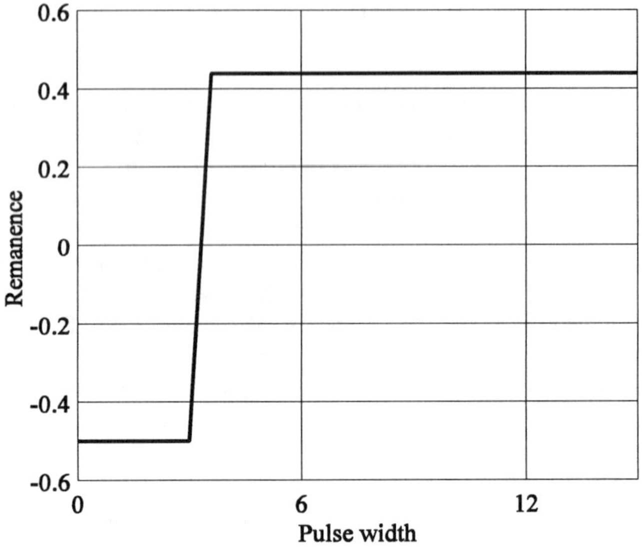

Figure 7.8 Pulse width dependence of the remanence.

width after the nucleation time.

The pulse height dependence of the remanence for the same pulse shown in Fig. 7.7 is illustrated in Fig. 7.9. It is seen that the remanence varies from $-S$ to $+S$, where in this case, S is 0.5. For this choice of material parameters, this curve is essentially the same as the major remanence loop for this material.

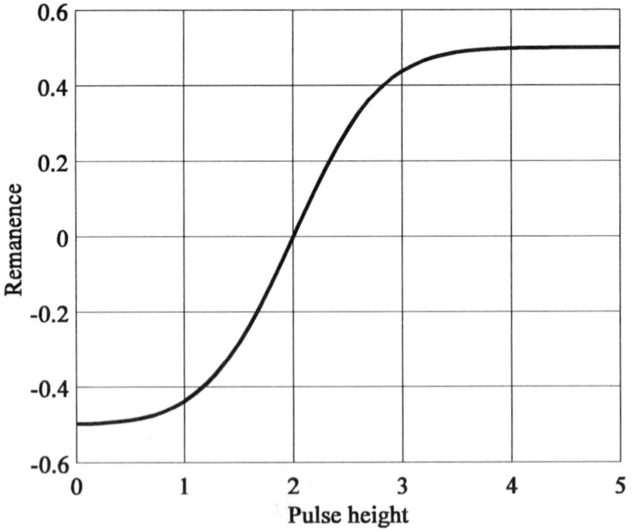

Figure 7.9 Pulse height dependence of the remanence .

7.5.3 Double-Pulse Simulation

We now assume that the medium is saturated in the "down" direction. At $t = 0$, a pulse in the "up" direction whose strength is H_1 and whose duration is t_{D1} is applied. This is followed by a second "up" pulse, at $t = t_1$, whose strength is H_2 and whose duration is t_{D2}. The initial condition is different when the two pulses are applied. For the first pulse the entire fourth quadrant of the Preisach plane had a Q of -1.

For the second pulse the fourth quadrant is divided vertically in two regions. To the left of the line $v = h_{max}$, $Q = 1$ at the end of the pulse if t_{D1} is greater than the sum of t_N and t_R. It will then accommodate to a value determined by the change in m_r. For shorter pulses the change in magnetization will be increased by the completion in the change in m_i. These two changes will have the opposite effect. To the right of the line $v = h_{max}$, the value of Q starts from -1 at the beginning of the pulse and will accommodate to a more positive value. In both cases the value of Q is computed from (7.17).

When the second pulse is applied, m_r will change immediately. The region to the left of the operative field will then start reversing as indicated by (7.22). In this case, the value of Q will increase only slightly, since only the hysterons that have accommodated into that region need to be reversed. The analysis is more complicated if the height of the second pulse is different from that of the first pulse.

For second pulse heights greater than the first pulse, the analysis is similar to that given for the first pulse. For smaller pulse heights, the Preisach plane is divided into three regions: the region to the left of the new operative field, that between the new operative field and the old one, and that to the right of the old operative field. The first region will have $Q = 1$ as long as the pulse is applied.; the second region will have a Q somewhat less than 1; and in the third region Q will have a value somewhat more than -1. The difference between the magnitude of these Q's and 1 is due to accommodation.

7.6 NOISE

The theory of Barkhausen noise in recording media has been studied extensively for recording media consisting of noninteracting hysterons. This noise occurs because the magnetization changes in discrete steps, and as a result, the magnetization curve is a staircase instead of a smooth curve, as shown in Fig. 7.10. A smooth curve would have no noise. Interaction between hysterons increases noise by reducing the number of independent magnetic states available to the system by the cooperative magnetization of otherwise independent hysterons [9]. The inclusion of interaction into this theory requires a physical model of the magnetizing process. In this section we will use an extended Preisach model that includes accommodation and noncongruency effects.

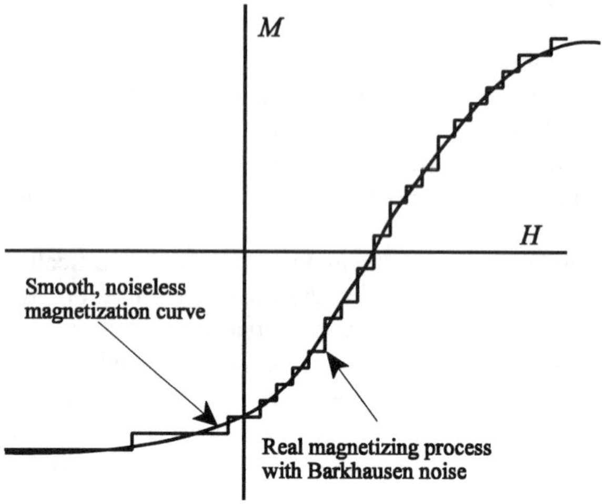

Figure 7.10 Staircase ascending major loop as a result of Barkhausen noise in the magnetizing process as contrasted to a smooth noiseless magnetization curve.

In addition to the other sources of noise in a recording system, Mallinson summarizes the theory of Barkhausen noise in noninteracting particulate recording media in his excellent summary article [10]. He shows that the noise power P_N of a fully saturated recorded bit, if all the hysterons in the medium are identical, is given by

$$P_N = 4\pi m^2 N w V^2 \frac{\delta(d + \delta/2)}{d^2(d + \delta)^2}, \tag{7.25}$$

where m is the dipole moment of each hysteron, N is the number of hysterons per unit volume, w is the track width, V is the head-to-medium velocity, δ is the coating thickness, and d is the head-to-medium spacing. This formula is derived with these assumptions:

The head efficiency is 100%.
The head is able to capture all the flux from the recorded bit.
The recording medium is very thin.
The head has one turn.
There is no gap loss.
The head is connected into a one-Ohm load.
The hysterons do not interact.

The assumption that the recording medium is thin implies that the magnetization is uniform throughout the thickness of the coating. This assumption, like the others, can easily be corrected.

The effect of hysteron interaction, on the other hand, requires some knowledge of how interaction affects the recording process. We will now examine the effect of medium thickness.

Let us define K to be the number of hysterons in a half-wavelength, $\lambda/2$. For thin coatings, say less than one-third of a wavelength, we can assume that the recording is uniform throughout the coating. For thicker coatings, however, the penetration depth of the recording into the coating is limited by the wavelength. We can adopt the following rule of thumb for K:

$$K = \begin{cases} Nw\lambda\delta\ , & \text{if } \delta \le \lambda/3 \\ Nw\lambda^2/3\ , & \text{if } \delta > \lambda/3 . \end{cases} \tag{7.26}$$

In the remaining equations in this section, we will assume that $\delta \le \lambda/3$, so that (7.25) may be written

$$P_N = 4\pi m^2 K V^2 \frac{d + \delta/2}{\lambda d^2 (d + \delta)} . \tag{7.27}$$

The noise power spectral density in a wave interval number Δk is given by

$$e_N^2(k) = \frac{4\pi m^2 K V^2}{\lambda\delta} |k| (1 - e^{-2|k|\delta}) e^{-2|k|d} \Delta k , \tag{7.28}$$

where k is given by

$$k = \frac{2\pi f}{V} . \tag{7.29}$$

For sine wave recording, the maximum possible signal power spectrum is given by

$$e_S^2(k) = \left[\frac{\pi m K V}{2\lambda\delta} (1 - e^{-|k|\delta}) e^{-|k|d} \right]^2 . \tag{7.30}$$

Thus, the maximum *signal-to-noise ratio SNR* is given by

$$SNR = \frac{e_S^2(k)}{e_N^2(k)} = \frac{K(1 - e^{-|k|\delta})^2}{\delta(1 - e^{-2|k|\delta})\Delta k} . \tag{7.31}$$

It is seen that the *SNR* is independent of hysteron moment or the head-to-medium velocity; however, this is only the *SNR* due to Barkhausen noise. The contribution to the *SNR* due to the remainder of the noise does vary with the hysteron moment. We shall modify these formulas by including the effect of interaction calculated by the CMH model with accommodation.

7.6.1 The Magnetization Model

A system of K noninteracting hysterons has 2^K possible states. Of these states, a magnetizing process starting at negative saturation and going to positive saturation traverses K of these states. As the field is increased, hysterons with the lowest critical field will switch first. The distribution of critical fields affects only the linearity of the magnetizing process. For example, if the distribution is Gaussian, the magnetization curve is an error function. If a system of interacting hysterons is describable by the classical Preisach model, then when the field is increased just enough to switch one hysteron, only that hysteron will switch. Therefore, the hysterons may switch in a different order depending on the history of the process, but there will be no effect on the number of available magnetization states. Thus, the main effects of this type of interaction are to modify the magnetization curve and to redistribute the magnetization noise, not the total noise power associated with the process; but the system will still traverse K states in going from negative saturation to positive saturation.

The effect of interaction normally increases the amount of noise by reducing the number of independent hysterons that can switch. That is, interaction can decrease the number of independent states when pairs of hysterons switch as a single unit. Hence, this additional noise will be referred to as *excess Barkhausen noise*. This effect may also modify the magnetization curve by redistributing the switching fields.

Two modifications have been made to the classical Preisach model which affect the number of available states: the *moving* model modification [11] and the *accommodation* model modification [3]. These two modifications have removed the *congruency property* limitation and the *deletion property* limitation, respectively, of the classical Preisach model. Although the product model [12] also removes the congruency property, it does not appear to be applicable to recording media.

7.6.2 The Effect of the Moving Model

We will assume that K, the total number of hysterons in the system, is the sum of the number of hysterons that switch independently, K_{ind}, and the number of groups of hysterons that switch cooperatively, K_{coop}; that is

$$K = K_{ind} + K_{coop} . \tag{7.32}$$

When the applied field is increased by ΔH, two regions are switched in the operative plane, as shown in Fig. 7.11. In this process, ΔK hysterons are switched; ΔK_{ind} of them are switched independently, and ΔK_{coop} are switched cooperatively. Thus,

$$\Delta K_{ind} = \int_I P(u,v)\,du\,dv, \tag{7.33}$$

and

$$\Delta K_{coop} = \int_{II} P(u,v)\,du\,dv, \tag{7.34}$$

where P is the Preisach function. Then,

$$\frac{\Delta K}{\Delta K_{ind}} = 1 + \frac{\Delta K_{coop}}{\Delta K_{ind}}. \tag{7.35}$$

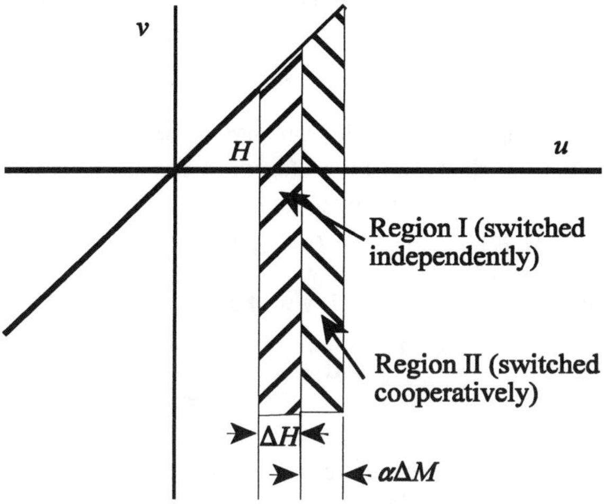

Figure 7.11 Regions of the operative plane that are switched when the applied field in increased from H to $H + \Delta H$.

For this change in the applied field, the number of hysterons that were independently switched is now given by

$$\Delta K_{ind} = \frac{\Delta K}{1 + \Delta K_{coop}/\Delta K_{ind}}. \tag{7.36}$$

For small changes in ΔH, the ratio of ΔK_{ind} to ΔK_{coop} is given by $\alpha\ \Delta M/\Delta H$. Thus, (7.36) can be rewritten

$$\Delta K_{ind} = \frac{\Delta K}{1 + \alpha\Delta M/\Delta H} = \frac{\Delta K}{1 + \alpha\chi}, \tag{7.37}$$

where χ is the susceptibility. It is seen that the number of independent states is normally smaller than the number of hysterons, since χ is positive and α is normally positive. If α were ever to be negative, the *SNR* in some cases could be greater than the case for noninteracting hysterons; however, this may be permitted, since there are many more states than just those traversed when the hysterons do not interact. Furthermore, since χ is a function of both the applied field and the magnetization, the decrease in the number of independent states depends on both the magnetization and the applied field. The total number of independent states is then obtained by integrating (7.37); that is,

$$K_{ind} = \int_{-\infty}^{\infty} \frac{dK/dH}{1 + \alpha\chi}\, dH. \tag{7.38}$$

It is seen that if α is zero, then the number of independent states is the same as the total number of hysterons.

We will decompose the susceptibility χ into a reversible component X_r and an irreversible component χ_i. Thus,

$$\chi = \chi_r + \chi_i . \tag{7.39}$$

This is a function of both the magnetic state and the applied field. In particular, if the reversible function, $f(H)$, can be factored, in the same way as in the CMH model, the reversible susceptibility is given by

$$\chi_r(M,H) = a_+(M)\, f(H) - a_-(M)\, f(-H). \tag{7.40}$$

For the simplified case, the DOK model, we have

$$a_+ = \frac{m_i + 1}{2} \quad \text{and} \quad a_- = \frac{m_i - 1}{2}, \tag{7.41}$$

where m_i is the normalized irreversible component of the magnetization. Otherwise, the a's are Preisach-like integrals.

The irreversible component of the susceptibility as computed by the Preisach model is

$$\chi_i(H,M) = \int_{-\infty}^{H+\alpha M} p(H + \alpha M, v)\, dv, \tag{7.42}$$

where p is the Preisach function. Substituting these equations into (7.38) gives the number of independent states as computed from the CMH model alone.

7.6.3 The Effect of the Accommodation Model

The accommodation model introduces a second type of cooperative effect that often occurs in hysteretic many-body problems. Not only does the Preisach function move in the plane, but also hysterons will change their position within the function. Whenever hysterons change their position, they may cross the line defined by the applied field H as shown in Fig. 7.12. If the change in magnetization causes the hysteron in the illustration to move from position 1 to position 2 in the hatched region, the hysteron will switch. This is an additional example of cooperative switching which further reduces the number of states available.

The degree of saturation of a region, Q, can be computed by the accommodation model as

$$\frac{dQ}{dH} = \frac{\beta(M_i - SM_S Q)}{S^2 M_S^2}\left|\frac{dM_i}{dH}\right|, \tag{7.43}$$

where β is the accommodation constant. We see that if β is zero, there will be no dilution of the magnetization in any region. In this case, the rate of change in the number of independent states is given by

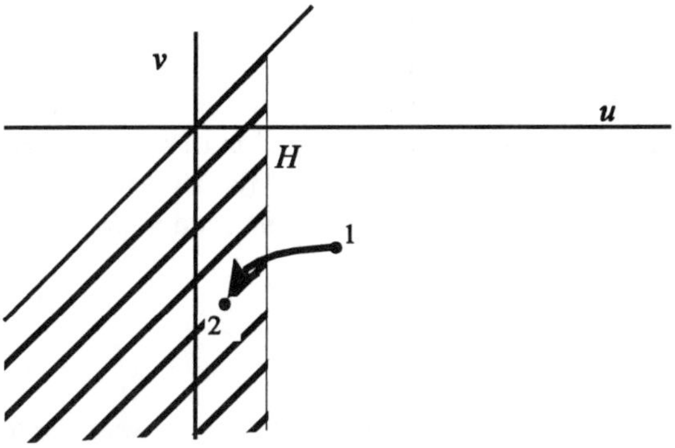

Figure 7.12 When a field is applied, the hatched region is magnetized in the positive direction. As a hysteron moves from position 1 to position 2, the local field becomes sufficient to magnetize it positively.

$$\frac{dK}{dH} = -K\frac{dQ}{dH} \int\int_{H<H+\alpha M} p(u,v)\,du\,dv. \qquad (7.44)$$

It is noted that this quantity is negative, since Q increases when the field increases and the Preisach integral is always a positive fraction. We can write this as

$$\frac{\Delta K}{K} = \frac{\beta(M_i - SM_S Q)}{S^2 M_S^2}\Delta M_i \int_{H<H+\alpha M} P(u,v)\,du\,dv. \qquad (7.45)$$

It is seen that if β is zero, there will be no change in the number of available states, hence, no excess Barkhausen noise. The region of integration is the region where the hysterons have a positive switching field that is smaller than the applied operative field. It is noted that the quantity on the right-hand side of (7.45) is less than 1, so that the number of states is again smaller. We see that if β is zero (i.e., there is no minor loop accommodation), there is no decrease in the number of independent hysterons due to this process. It is noted that although the major loop does not accommodate, it is still susceptible to this type of excess Barkhausen noise.

In a system with both motion and accommodation, the excess noise is the sum of the two effects. Furthermore, the two effects interact: Any accommodation produces a change in magnetization, which moves the Preisach function and results in a loss of independent states due to motion; also any motion changes the magnetization, which in turn causes accommodation. For completeness, the effect of reversible magnetization must be included into the accommodation calculations.

It is noted that the cooperative effect is not the same for all magnetizations. In particular, the moving model produces less excess noise when the susceptibility is small, such as the case of near saturation. The accommodation model also produces less excess noise near saturation, since the accommodation model is driven by the change in magnetization.

The analysis above was carried out for an increasing applied field. For an applied field decreasing from positive saturation to negative saturation, the signs of *dM/dH* must be changed. In this case, the overall effect is still the same: Both the moving model and the accommodation model decrease the number of independent states.

7.7 MAGNETOSTRICTION

Highly magnetostrictive media, such as Terfenol-D, are useful for transducer applications [13], but are also hysteretic. Their usefulness as linear actuators is limited to a small fraction of their capability unless they can be accurately controlled [14]. The first step in controlling these materials is to develop an accurate, efficient model. Modeling of this material has been extensively discussed in earlier work [15] and [16,17]. Here, we modify Preisach models with state-dependent reversible magnetization to model magnetostrictive behavior [18].

Both the magnetization and the strain of a magnetostrictive material are hysteretic when viewed as a function of applied field. Two moving Preisach models—the DOK model [19] with magnetization-dependent reversible magnetization, and the more accurate CMH model [20] with state-dependent reversible magnetization—can accurately characterize the magnetization of some media. In this section, the DOK model is modified to also characterize the strain of magnetostrictive material.

Figure 7.13 shows a typical plot of measured strain versus applied field for the particular magnetostrictive material Terfenol-D [21]. For this material, strain is an

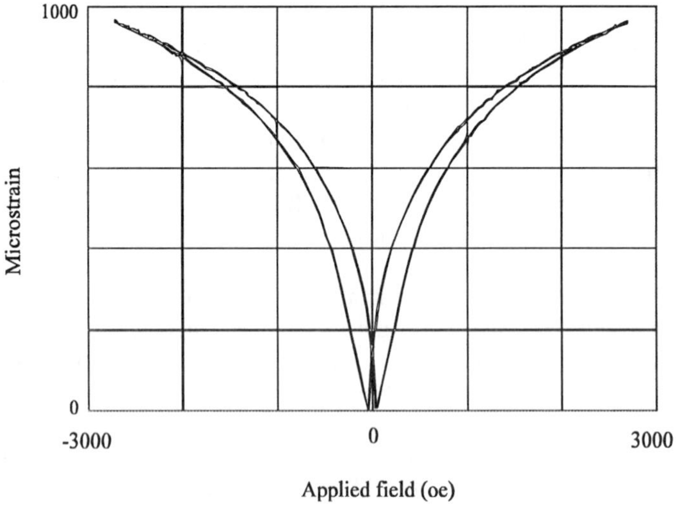

Figure. 7.13 Measured strain vs applied field for Terfenol-D (courtesy of J. E. Ostensen and D. C. Jiles).

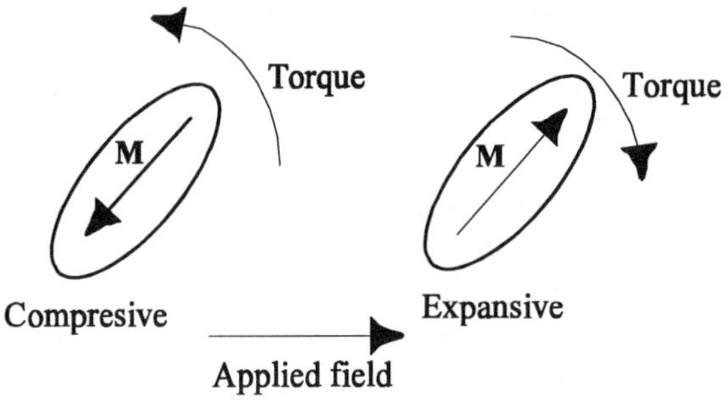

Figure 7.14 Effect of an applied field on an acicular particle.

expansive, even function of the applied field; that is, it elongates in the presence of a field.

To model this behavior, we will assume that the medium consists of hysterons, which are either particles or grains whose shape may be acicular or platelet. Because of the anisotropy of the hysterons, if their axes are not perfectly aligned with the applied field, the medium will not have unity squareness. When a field is applied to this medium, a torque is applied to each hysteron, which in turn applies a stress to the medium, since the hysterons have shape anisotropy.

The torque, and consequently the stress, depends on the direction of the magnetization along the hysteron's easy axis, and thus is state dependent. As illustrated in Fig. 7.14, if the applied field makes an obtuse angle with the magnetization, which we will call the "negative magnetization state," the stress is compressive for acicular hysterons. If it makes an acute angle, which we will call the "positive magnetization state," the stress is expansive. This set of definitions implies that if the medium is demagnetized, the stress field is zero. However, if the material is magnetized, it is not zero.

Change in magnetization is due to the rotation effected by the torque supplied by the applied field. This rotation from the hysteron's easy axis is opposed by the variation in the demagnetizing field for hysterons with shape anisotropy. When the applied field is removed, the magnetization will return to the easy axis. Thus, the rotational energy supplied by the applied field is returned when the field is removed.

An applied field also produces a torque on the hysteron, which attempts to rotate it in the same direction that the magnetization is rotated. In the case of the magnetization, there is a restoring torque due to the hysteron's shape or due to the magnetocrystalline anisotropy of the particle/grain. In the case of magnetostriction, the rotation is opposed by the binder that holds the material together. Assuming that the magnetization of the hysteron is constrained to its long axis, then in both cases, a certain amount of rotation produces the same fractional increase in magnetization as the fractional increase in length (strain).

In both the DOK and the CMH models, the reversible component of the magnetization M_r is given by

$$M_r = SM_S[a_+ f(H) - a_- f(-H)], \qquad (7.46)$$

where S is the squareness, determined by the angular distribution of the hysterons, M_S is the saturation magnetization, and $f(H)$ is the normalized reversible component of the magnetization when the hysteron is in its positive state. The squareness is the ratio of the maximum remanence to the saturation magnetization. The function $f(H)$ is essentially determined by the hysteron's anisotropy and is a monotonic function that approaches unity asymptotically as H becomes large. The difference between the DOK and CMH models is in the method of calculating the a's. In the DOK model, the a's are given by

$$a_+ = \frac{M_i + M_S}{2M_S} \quad \text{and} \quad a_- = \frac{M_i - M_S}{2M_S}, \qquad (7.47)$$

where M_i is the irreversible magnetization. That is, the DOK model is magnetization dependent. The CMH model, which is state dependent, uses a more complex expression for the a's.

Since the functional variation of the reversible magnetization and the magnetostriction are the same in a given state, we will use the same function for both; however, for a magnetostrictive material, the stress and the reversible magnetization have the opposite effect when the magnetization is in the negative state. Therefore, we will describe the stress T by

$$T(H,M_i) = K\,[a_+(M_i)\, f(\nu H) + a_-(M_i)\, f(-\nu H)], \qquad (7.48)$$

where the a's are given by (7.47). Since $f(H)$ approaches unity as H becomes large, we have to introduce the factor K for length normalization. Similarly, we introduce the factor ν as the ratio of magnetostrictive susceptibility to magnetic susceptibility. The latter factor is determined by the relative effectiveness of the magnetic anisotropy and the binding forces that try to keep the hysteron oriented in a particular way. The constant K is positive for acicular hysterons and negative for platelet hysterons. This produces the desired properties that T is zero if H is zero or if $a_- = a_+$ (zero magnetization in the DOK model). Since the DOK model does not differentiate between the demagnetized states and the CMH model does, the two models will behave differently for demagnetized media.

In the present simulation we assume that the magnetic material is represented by the DOK model and that the moving constant is zero. We will further assume that the Preisach function is Gaussian and that the reversible field variation is given by

$$f(H) = 1 - e^{-\xi H}, \qquad (7.49)$$

where ξ is the normalized zero field susceptibility. Starting with a *dc*-demagnetized specimen, the irreversible component of the magnetization is given by

$$M_i = SM_S \operatorname{erf}\left(\frac{H-H_{\text{rem}}}{\sigma}\right), \tag{7.50}$$

where H_{rem} is the remanent coercivity and σ is the standard deviation of the switching field.

Figure 7.15 plots both T and T/H as the field is increased from the demagnetized state (lower curve) and then reduced back to zero. The T–H plot could be compared to the first quadrant of Fig. 7.13, if the stress–strain relationship were linear. A nonlinear relationship would further modify this curve. Note that since $f(0)$ is zero, T is zero when the applied field is zero. Furthermore, since in the DOK model $a_+(0) = -a_-(0)$, we have

$$\left.\frac{dT(H,0)}{dH}\right|_{H=0} = K[a_+(0) + a_-(0)] = 0, \tag{7.51}$$

and the slope of T is zero. Additionally, since

$$a_+(M_i) = -a_-(-M_i), \tag{7.52}$$

starting from a demagnetized state, and since the error function is an odd function, T is an even function of H, which is consistent with what is observed in Fig. 7.14. Since the curve has even symmetry, it is plotted for positive fields only.

As the field is increased, the ratio T/H increases at first. Since $f(H)$ is positive and $f(-H)$ is negative, the increase in T/H is due to increase in a_+ at the expense of

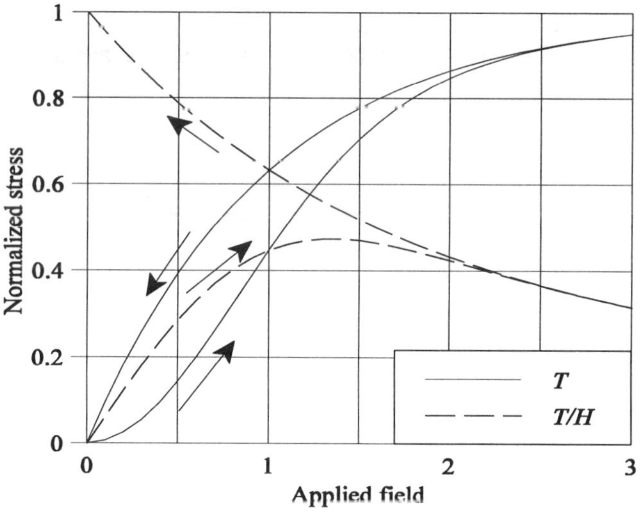

Figure 7.15 Calculated stress relationship using the DOK model.

a_-. The ratio then decreases as a result of the saturation of the numerator and the continued increase in the denominator. When the field decreases, the ratio increases, since the denominator decreases faster than the numerator.

It is seen that this model generates hysteresis close to that seen in these materials, with a few exceptions. The stress in this model is zero in the absence of an applied field, contrary to the measurements. If the stress were a function of the operative field, $h = H + \alpha M$, instead of the applied field H, there would be a remanent stress in the material. Future versions of the model will be based on the operative field rather than the applied field. Furthermore, the slope of T at $H = 0$ is not observed to be zero, as the model predicts. This reflects the model's assumption of starting with a completely demagnetized sample, while the measured data was taken on a sample that was not demagnetized. It may also be the result of a nonlinear stress–strain relationship.

The actual strain depends on the stress-strain relationship of the medium and the load placed on the transducer. For a linear stress-strain relationship, the medium strain is found by Young's modulus times the stress. For a hysteretic relationship, the strain could be calculated by a second Preisach model. In the latter case there might be some residual stress at the conclusion of this process. If found to be necessary, this relatively simple modification of the model requires little additional computing time and introduces only one additional arbitrary constant.

This model has the capability of predicting minor loop behavior as seen in Fig. 7.16. In this case, the material is assumed to be single-quadrant material, so that there is no change in state as the applied field is decreased. Thus, the a_+ and a_- do not change, and the shape of the curve is determined entirely by the shape of $f(H)$.

Examination of Fig. 7.17 illustrates the effect of varying ν, as defined in (7.48), for the values of 0.15, 0.6, 1.35, and 2.4. It is seen that as it increases, both the slope and height of the hysteresis loop increase for a given field. These curves are normalized in Fig. 7.18 so that the subtle changes in shape are more easily

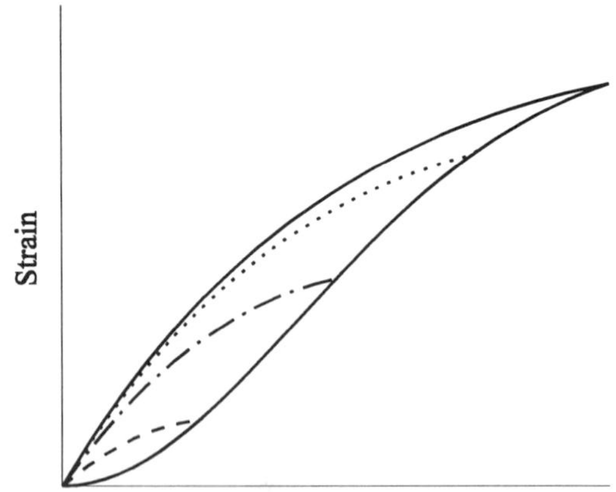

Figure 7.16 Minor loops calculated by the magnetostriction model.

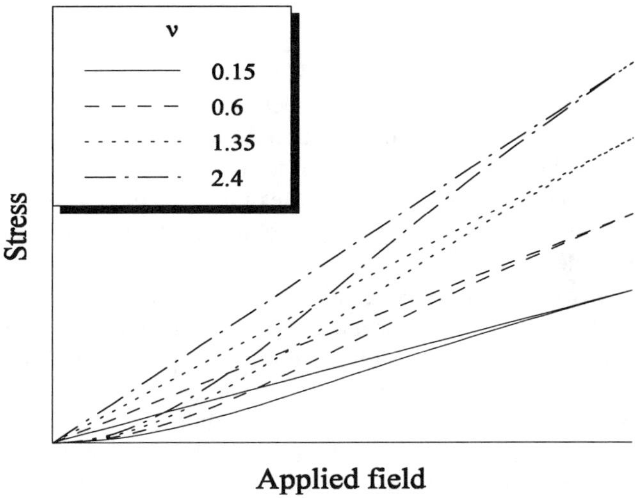

Figure 7.17 The effect of ν on the model's magnetostriction behavior.

compared. As ν is increased, the curvature of the stress increases as the field is decreased. As the field is increased, the effect of increasing ν is to emphasize the decrease in slope at larger fields making the curve more "S" shaped.

A Preisach-type model has been presented for modeling magnetostriction of particulate or granular media. The stress-applied field relationship generated by this model displays the hysteresis observed in magnetostrictive media, even though only a very simplified DOK model was used. Again, to validate the model, one must identify the parameters completely, to upgrade to a CMH model, and to accurately model the stress–strain relationship in the context of the overall model.

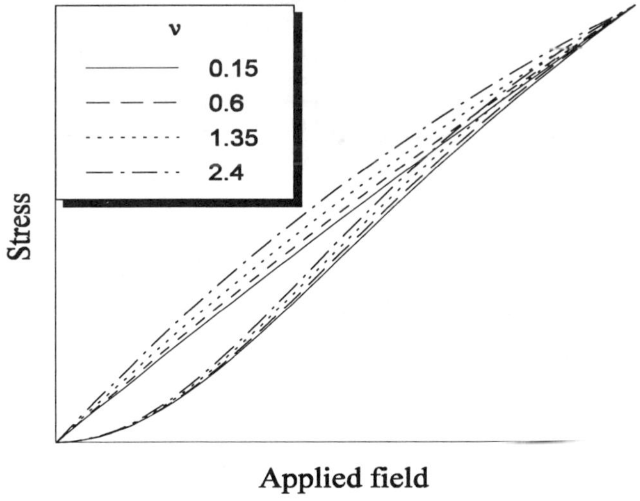

Figure 7.18 Normalized magnetostriction behavior for different values of ν.

7.8 THE INVERSE PROBLEM

An important application of hysteresis modeling is the inverse problem, in which an appropriate circuit is obtained to condition the input to the hysteretic transducer so that the overall circuit does not appear to have hysteresis. Visone et al. have shown [22] that the stop model is the inverse of the play model (see Appendix A). We will now show how to obtain the inverse of the differential equation form of the Preisach model.

We visualize the Preisach model as consisting of three components, as shown in Fig. 7.19: a differentiation, a susceptibility computer, and an integrator. The time derivative of the operative field is obtained as the output of a differentiation. For the Gaussian Preisach density function, the susceptibility χ is computed using (2.22), to obtain the irreversible component. In addition, we must obtain a suitable model for the reversible component—for example, by means of (3.12). The choice of the a's is determined by whether one uses the magnetization-dependent or the state-dependent model for the reversible magnetization.

The history of the process is maintained by a stack in the box labeled "Compute χ." The time derivative of the operative field is then multiplied by the susceptibility to obtain the time derivative of the magnetization, which is in turn integrated to obtain the magnetization. Other features could be added to this model, such as accommodation and aftereffect; however, we will not do so, since their presence would complicate this picture.

Since the susceptibility is a scalar, the inverse of this transducer involves replacing the multiplication by χ with division by χ, as shown in Fig. 7.20. The reciprocal susceptibility is computed the same way that the susceptibility is computed, and the same stack is used to maintain the magnetizing history. Since the output of this circuit is integrated, it will be less sensitive to noise. The errors associated with this inverse are associated with the approximation of the model to the real system. In particular, the parameters of the system may change with time

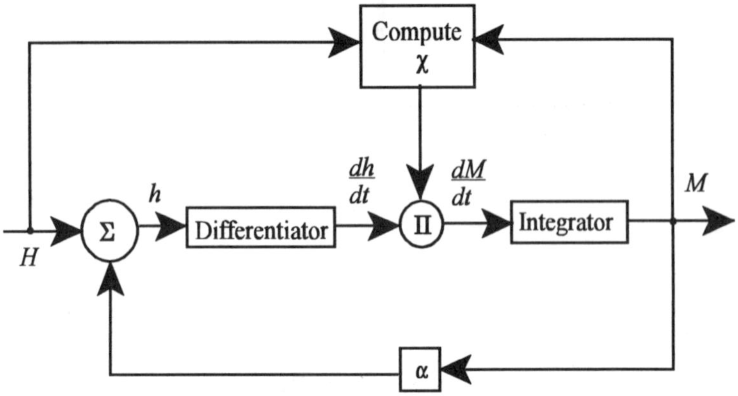

Figure 7.19 Block diagram of the differential Preisach model.

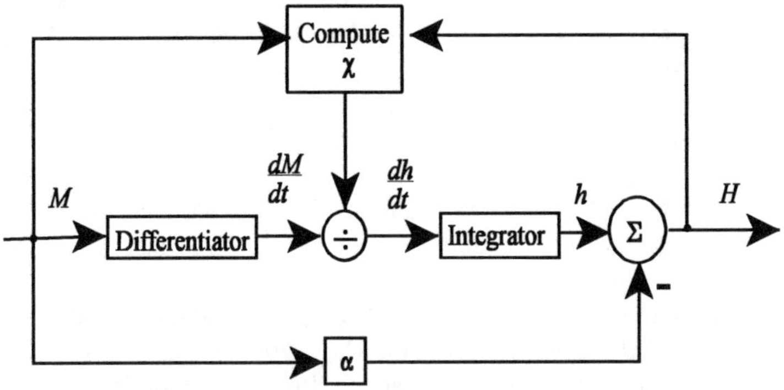

Figure 7.20 Block diagram of the inverse differential Preisach model.

as the temperature of the transducer changes, and the model may not track them correctly. Other errors are associated with approximating the critical field Preisach density and with approximating the reversible variation.

This model has a self-correcting property. Whenever the applied field becomes large, the material and the inverse model go to a unique state, the saturation state. Furthermore, the errors associated with the improper registration of a corner of the history staircase are deleted whenever that corner is deleted by the applied field. This inverse is both a left inverse and a right inverse.

7.9 CONCLUSIONS

The classical Preisach model is able to describe hysteresis but is limited by the congruency property and the deletion property. These properties are not found in magnetic materials, and so the model must be modified accordingly. Furthermore, the model is a scalar one, and real magnetizing processes are vector ones. In earlier chapters we showed how physical arguments could be used to modify these properties. The results were accurate models that had relatively few parameters and gave some insight into the magnetizing process.

In this chapter we showed how to introduce dynamics into the rate-independent Preisach model. One can also obtain a robust model that is capable of describing far more than just the magnetization characteristics of the material. One example of such an extension of the model is the magnetostriction model. In addition, since the Preisach model possesses an inverse, it can be used if desired to modify the input so that the resulting transducer appears to have no hysteresis.

REFERENCES

[1] G. Bertotti, "Dynamic generalization of the scalar Preisach model of hysteresis," *IEEE Trans. Magn.*, **MAG-28**, September 1992, pp. 2599–2601.

[2] E. Della Torre, "An analysis of the frequency response of the magnetic recording process," *IEEE Trans. Audio Electroacoust.,* **AE-13**, May–June 1965, pp. 61–65.

[3] W. D. Doyle, L. Varga, L. He, and P. J. Flanders, "Reptation and viscosity in particulate recording media in the time-limited switching regime," *J. Appl. Phys.*, **75**, May 1994, pp. 5547–5549.

[4] P. J. Flanders, W. D. Doyle, and L. Varga, "Magnetization reversal in magnetic tapes with sequential field pulses," *IEEE Trans. Magn.*, **MAG-30**, November 1994, pp. 4089–4091.

[5] Y. D. Yan and E. Della Torre, "Particle interaction in numerical micromagnetic modeling," *J. Appl. Phys.*, **67(9)**, May 1990, pp. 5370–5372.

[6] E. Della Torre, J. Oti, and G. Kádár, "Preisach modeling and reversible magnetization," *IEEE Trans. Magn.*, **MAG-26**, November 1990, pp. 3052–3058.

[7] E. Della Torre, "Dynamics in the Preisach accommodation model," *IEEE Trans. Magn.*, **MAG-31**, November 1995, pp. 3799–3801.

[8] E. Della Torre and F. Vajda, "Parameter identification of the complete-moving hysteresis model using major loop data," *IEEE Trans Magn.*, **MAG-30**, November 1994, pp. 4987–5000.

[9] E. Della Torre, "Effect of particle interaction on recording noise," *Physica B*, **223**, 1997, pp. 337–341.

[10] J. C. Mallinson, in *Magnetic Recording, Vol. I,* C. D. Mee and E. D. Daniels, eds. McGraw-Hill: New York, 1987, pp. 337–375.

[11] E. Della Torre, "Effect of interaction on the magnetization of single domain particles," *IEEE Trans. Audio Electroacoust.,* **AE-14**, June 1966, pp. 86–93.

[12] G. Kádár, "On the Preisach function of ferromagnetic hysteresis," *J. Appl. Phys.*, **61**, April 1987, pp. 4013–4015.

[13] M. B. Moffet, A. E. Clarke, M. Wun-Fogle, J. Linberg, J. P. Teter, and E. A. McLaughlin, "Characterization of Terfenol-D for magnetostriction transducers," *J. Acoust. Soc. Am.*, **89(3)**, 1991, pp. 1448-1455.

[14] F. T. Calkins, and A. B. Flatau, "Transducer based measurements of Terfenol-D material properties," *SPIE 1996 Proc.: Smart Structures and Integrated Systems*, **2717**, 1996, pp. 709–719.

[15] J. B. Restorff, H. P. Savage, A. E. Clark, and M. Wun-Fogle, "Preisach modeling of hysteresis in Terfenol," *J. Appl. Phys.,* **67(9)**, May 1990, pp. 5016–5018.

[16] I. D. Mayergoyz, *Mathematical Models of Hysteresis*, Springer-Verlag: New York, 1990, pp. 122–129.

[17] A. Adly and I. D. Mayergoyz, "Magnetostriction simulation by using anisotropic vector Preisach models," *IEEE Trans. Magn.*, **MAG-32**, November 1996, pp. 4147–4149.

[18] E. Della Torre and A. Reimers, "A Preisach-type magnetostriction model for magnetic media," *IEEE Trans. Magn.,* **MAG-33**, Sepember 1997, pp. 3967–3999.

[19] E. Della Torre, J. Oti, and G. Kádár, "Preisach modeling and reversible magnetization," *IEEE Trans. Magn.*, **MAG-26**, November 1990, pp. 3052–3058.

[20] E. Della Torre and F. Vajda, "Parameter identification of the complete-moving-hysteresis model using major loop data," *IEEE Trans. Magn.*, **MAG-30**, November 1994, pp. 4987–5000.

[21] J. E. Ostenson and D. C. Jiles, Ames Laboratory, Iowa State University, private communication.

[22] C. Miano, C. Serpico, and C. Visone, "A new model of magnetic hysteresis, based on stop hysterons: an application to the magnetic field diffusion," *IEEE Trans. Magn.*, **MAG-32**, May 1996, pp. 1132–1135.

APPENDIX A

THE PLAY AND STOP MODELS

The play model* is another method of handling certain types of hysteresis. Hysteresis observed in mechanical systems, such as gear trains, is called *backlash*. In particular in a gear train, there is a range of input, called the *dead zone,* which produces no change in output. When that range has been exceeded, the change in output is directly proportional to the change in input until the direction of the input is changed. At that point, the gear train reenters the dead zone. The ratio of the change in output to the change in input is called the *gear ratio*.

A graphical description of this behavior is shown in Fig. A.1. The sloping line going through point a is followed only whenever the input is increasing. The slope of this line is given by the gear ratio. Similarly, the line going through point b also has a slope given by the gear ratio and is followed only whenever the input is decreasing. These sloping lines will be referred to as the *bounding lines*. The region between these lines is the dead zone. At point c, if the input increases, the output follows the line going through point a, but if the input decreases, it follows the horizontal line into the dead zone. This is an example of a *branching* in this model. The horizontal line can be traversed in either direction, and one follows this line as long as one is in the dead zone. A reversible component in this model can be introduced by replacing the horizontal lines in the dead zone with curves of finite

*M. A. Krasnosel'skii and A. V. Pokrovskii, *Systems with Hysteresis*, Springer–Verlag: Berlin, 1989.

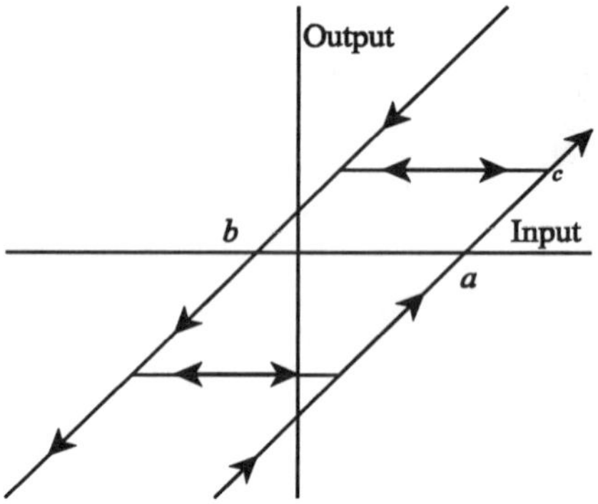

Figure A.1 Illustration of the play model.

slope.

The output of this model can be of any value and does not change as long as the input has a range of values defined by the width of the dead zone. Similarly for a given input, the output can have a range of values defined by the dead zone. The particular value of output for a given input depends on the history, so this system exhibits hysteresis. It is noted that the output of this model does not saturate as in the Preisach model. Thus, to use this model to characterize magnetic hysteresis, the output of the model must be fed into a saturating nonlinearity, as shown in Fig. A.2. T his cumbersome addition to the model limits its usefulness, especially when

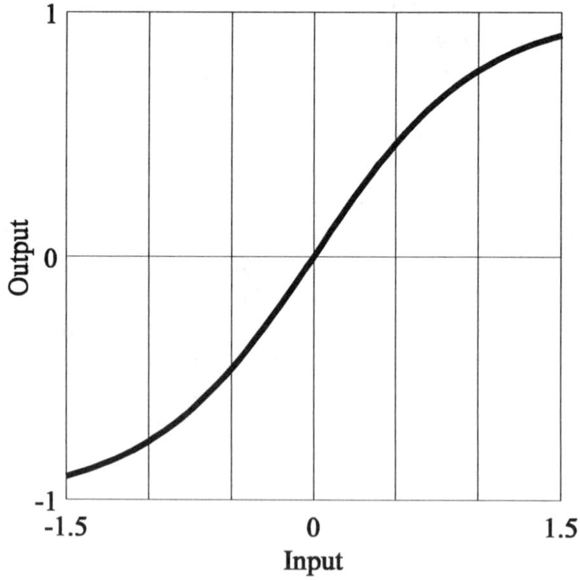

Figure A.2 Saturating nonlinearity.

trying to relate the model parameters to physical quantities.

The inverse of this model, the stop model, is a similar model where the slope of each of the lines in the inverse model are the reciprocal of the slope in the play model. In the dead zone, the slope of the curves would be infinite if there were no reversible component leading to discontinuities in the behavior. This is the same problem that the inverse of the Preisach model would have when the material is in saturation.

APPENDIX **B**

THE LOG-NORMAL DISTRIBUTION

The log-normal distribution is a modification of the normal (Gaussian) distribution for random variables that are constrained to be positive. We will define a log-normal distribution function by:

$$f(x) = A\exp\left\{-\left[\frac{\ln(x/b)}{2c}\right]^2\right\}, \tag{B.1}$$

where b and c are positive. A plot of this function is shown in Fig. B.1 for $A = 1$, $b = 1$ and three values of c.

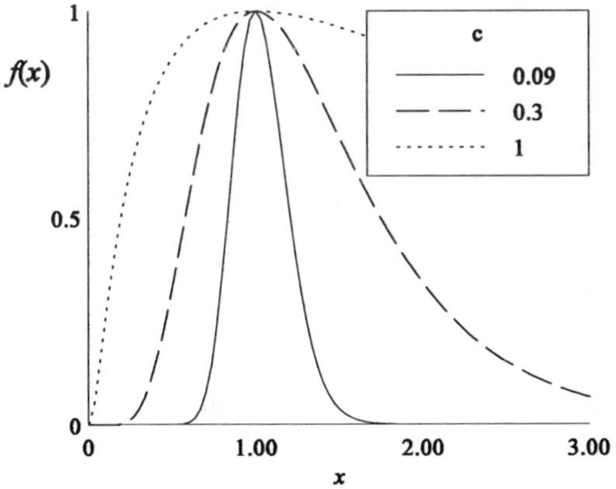

Figure B.1 Log-normal distribution.

The moment generating function *MGF* for this distribution is

$$MGF = \int_0^\infty f(x)\,dx = \int_0^\infty x^n f(x)\,dx = A\int_0^\infty x^n \exp\left\{-\left[\frac{\ln(x/b)}{2c}\right]^2\right\}dx. \tag{B.2}$$

If we make the substitution

$$u = \frac{\ln(x/b)}{2c}, \tag{B.3}$$

then

$$x = be^{2cu} \quad \text{and} \quad dx = 2bce^{2cu}du. \tag{B.4}$$

Thus, since

$$\int_{-\infty}^\infty \exp[-(x-c)^2]\,dx = \sqrt{\pi}, \tag{B. 5}$$

(B.2) becomes

$$\begin{aligned} MGF &= \int_{-\infty}^\infty 2Acb^{n+1}e^{-u^2+2cu(n+1)}du \\ &= 2Acb^{n+1}e^{c^2(n+1)^2}\int_{-\infty}^\infty \exp\{-[u-c(n+1)]^2\}du \\ &= 2\sqrt{\pi}Acb^{n+1}e^{c^2(n+1)^2}. \end{aligned} \tag{B.6}$$

We select A by normalizing the distribution, that is, setting the *MGF* to 1 for $n = 0$. Then

$$A = \frac{e^{-c^2}}{2bc\sqrt{\pi}}. \tag{B.7}$$

Therefore,

$$f(x) = \frac{e^{-c^2}\exp\left\{-\left[\frac{\ln(x/b)}{2c}\right]^2\right\}}{2bc\sqrt{\pi}}, \tag{B.8}$$

and the *MGF* becomes

$$MGF = b^n e^{c^2(n^2+2n)}. \tag{B.9}$$

The expected value of this distribution is obtained letting $n = 1$ in (B.9), so that

$$\langle x\rangle = be^{3c^2}, \tag{B.10}$$

and the expected value of x^2 is obtained by setting $n = 2$ in (B.9), so that

$$\langle x^2\rangle = b^2e^{8c^2}. \tag{B.11}$$

Then variance of $f(x)$, that is, the square of its standard deviation, is given by

$$\sigma^2 = \langle x^2 \rangle - \langle x \rangle^2 = b^2\left(e^{8c^2} - e^{6c^2}\right) = 2b^2 e^{7c^2}\sinh(c^2). \qquad \text{(B.12)}$$

We can now show that the log-normal distribution reduces to the Gaussian distribution if the standard deviation is small compared to the mean, that is, b is large and c is small. We note that for small values of c we have

$$\langle x \rangle \approx b \quad \text{and} \quad \sigma \approx 2bc. \qquad \text{(B.13)}$$

Note that if x is small compared to b, then $\ln(x/b)$ is approximately $(x/b - 1)$. Under these conditions, the distribution reduces to

$$f(x) \approx \frac{1}{\sigma\sqrt{2\pi}}\exp\left[-\frac{1}{2}\frac{(x - \langle x \rangle)^2}{\sigma^2}\right], \qquad \text{(B.14)}$$

which is a standard Gaussian distribution.

APPENDIX C

DEFINITIONS

Term	Symbol	Comment
Accommodation constant	β	Describes motion of hysterons in Preisach plane when M changes
Apparent reversible		Irreversible magnetization that has a unique state in zero field
Applied field	H_A	Applied magnetic field due to external sources
Applied field energy	W_{HA}	Energy due to interaction of the magnetization with an applied field
Coercivity	H_c	Field required to reduce the saturation magnetization to zero
Critical field	H_k	Field required to switch a noninteracting hysteron
Critical field expectation	$\overline{h_k}$	Average of hysteron switching operative fields

Cubic anisotropy energy	W_{cubic}	Magnetocrystalline energy of a cubic crystal
Demagnetization factor	D	Ratio of demagnetizing field to magnetization of a material
Demagnetizing field	H_D	Magnetic field due to the material's magnetization
Exchange constant	A	Exchange integral density for a simple cubic crystal equal to JS^2/a
Exchange integral	J	Exchange energy between adjacent atoms
Exchange energy	W_{ex}	Total exchange energy in a crystal
Hysteron		A minimum unit of magnetization with two stable states
Interaction field	H_i	Field due the magnetization of other hysterons
Irreversible magnetization	M_i	Component of the magnetization that changes irreversibly
Lattice spacing	a	Distance between adjacent atoms
Locally reversible magnetization	M_r	State-dependent reversible magnetization
Magnetic state	$Q(u, v)$	Magnetization state of a hysteron with critical fields u and v
Magnetization	M	Total magnetization per unit volume (the sum of the irreversible and locally reversible magnetization)
Moving constant	α	Ratio of expected value of interaction field to M
Negative critical field	v	Field required to switch a hysteron into the negative state
Positive critical field	u	Field required to switch a hysteron into the positive state

Preisach function	$P(u, v)$	Density function of hysterons with critical fields u and v
Remanence	M_0	Magnetization when H is zero
Remanent coercivity	H_{RC}	Field required to make the remanence zero
Saturation magnetization	M_S	Maximum magnetization for given material
Squareness	S	Ratio of maximum remanence to M_S
Susceptibility	χ	dM/dH, a subscript may be added to identify type of magnetization
Uniaxial anisotropy constant	K_u	Difference in energy density between easy axis and hard plane
Uniaxial anisotropy energy	W_u	Magnetocrystalline energy of a crystal with uniaxial symmetry
Wall width	l_w	Width of a domain wall, classically equal to $\pi\sqrt{A/K_u}$
Virgin magnetization curve	M_V	Magnetization curve for an *ac*-demagnetized specimen
Zeeman energy	W_H	Energy of a hysteron in a magnetic field

INDEX

ABOUT THE AUTHOR

Edward Della Torre received the B. E. E. degree from Brooklyn Polytechnic Institute in 1954, the M. Sc. in electrical engineering from Princeton University in 1956, the M. Sc. in physics from Rutgers University in 1961, and the D. E. Sc. from Columbia University in 1964. He has taught at Rutgers University, McMaster University, and Wayne State University, and he chaired the Electrical and Computer Engineering Departments at the latter two universities. He was a member of the technical staff at the Bell Telephone Laboratories in Murray Hill, NJ. He is Professor of Engineering and Applied Science at The George Washington University.

Dr. Della Torre has made fundamental contributions to the modeling of magnetic materials. A proponent of using physical principles to guide the development of models of magnetic materials, he developed the moving model, the state-dependent reversible magnetization model, the accommodation model, and the simplified vector model. Currently he is working on the Preisach-Arrhenius model for magnetic aftereffect, which determines the lifetime of magnetization.

A Fellow of both the IEEE and the American Physical Society, Dr. Della Torre is currently president of the Magnetics Society. He has served the Magnetics Society in many capacities including chairing several INTERMAG Conferences. He is the coauthor of *The Electromagnetic Field* with Charles V. Longo and *Magnetic Bubbles* with Andrew H. Bobeck. A member of Eta Kappa Nu, Tau Beta Pi and Sigma Xi, he is the author of almost 200 technical papers in refereed journals and has presented over 150 papers at technical conferences. He holds 18 patents.